FORTSCHRITTE DER PHYSIK

PROGRESS OF PHYSICS

Volume 32 · 1984

AKADEMIE-VERLAG · BERLIN

ISSN 0015-8208 Fortschr. Phys., Berlin **32** (1984)

The journal "Fortschritte der Physik/Progress of Physics" is devoted to the theoretical and experimental study of the fundamental constituents of matter and their interactions, so to elementary particle physics, classical and quantum field theory, theory of gravitation, thermodynamics and statistics, nuclear physics, laser physics, and plasma physics.
The articles should have in general review character.

Manuscripts should be addressed to the Editor-in-Chief of the journal

Prof. Dr. F. Kaschluhn
Sektion Physik der Humboldt-Universität zu Berlin
DDR-1086 Berlin, PSF 1297

or to editors or to members of the Advisory Board.

Terms of subscription for the journal "Fortschritte der Physik/Progress of Physics"

Orders can be sent

— *in the GDR:* to Deutsche Post, Zentralvertrieb des PZV (B), 7930 Herzberg/Elster or to the Akademie-Verlag, DDR-1086 Berlin, Leipziger Str. 3—4, PF-Nr. 1233;
— *in the other socialist countries:* to a book-shop for foreign language literature or to the competent news-distributing agency;
— *in the FRG and Berlin (West):* to a book-shop or to the wholesale distributing agency Kunst und Wissen, Erich Bieber OHG, Wilhelmstr. 4—6, D-7000 Stuttgart 1;
— *in the other Western European Countries:* to Kunst und Wissen, Erich Bieber GmbH, Dufourstr. 51, CH-8008 Zürich;
— *in other countries:* to the international book- and journal-selling trade, to Buchexport, Volkseigener Außenhandelsbetrieb der DDR, Postfach 160, DDR-7010 Leipzig, or to the Akademie-Verlag, DDR-1086 Berlin, Leipziger Str. 3—4, PF-Nr. 1233.

Zeitschrift „Fortschritte der Physik"

Herausgeber: Prof. Dr. F. Kaschluhn, Prof. A. Lösche, Prof. Dr. R. Rompe.

Verlag: Akademie-Verlag, DDR-1086 Berlin, Leipziger Str. 3—4;
Fernruf: 2236221 und 2236229; Telex-Nr.: 114420; Bank: Staatsbank der DDR, Berlin, Konto-Nr.: 6836-26-20712.

Chefredakteur: Dr. Lutz Rothkirch.
Anschrift der Redaktion: Sektion Physik der Humboldt-Universität zu Berlin, DDR-1040 Berlin, Hessische Straße 2.

Veröffentlicht unter der Lizenznummer 1324 des Presseamtes beim Vorsitzenden des Ministerrates der Deutschen Demokratischen Republik.

Gesamtherstellung: VEB Druckhaus „Maxim Gorki",
DDR-7400 Altenburg, Carl-von-Ossietzky-Straße 30/31.

Erscheinungsweise: Die Zeitschrift „Fortschritte der Physik" erscheint monatlich. Die 12 Hefte eines Jahres bilden einen Band. Bezugspreis je Band 216,— M zuzüglich Versandspesen. Preis je Heft 18,— M.

Bestellnummer dieses Heftes: 1027/32.

AN (EDV) 57618

Contents of Volume 32

Number 1

Number 2

Number 3

Number 4

Number 5

Number 6

Number 7

Number 8

Number 9

Number 10

Number 11

Number 12

FORTSCHRITTE DER PHYSIK

Volume 32 · 1984
Number 1

PROGRESS OF PHYSICS

CONTENTS:

AKADEMIE-VERLAG · BERLIN

ISSN 0015-8208 Fortschr. Phys., Berlin 32 (1984) 1, 1—24 EVP 10,— M

Instructions to Authors

1. Only papers not published and not submitted for publication elsewhere will be accepted.
2. Manuscripts should be submitted in English, with an abstract in English. Two copies are desired.
3. Manuscripts should be no less than 30 and preferably no more than about 100 pages in lenght.
4. All manuscripts should be typewritten on one side only, double-spaced and with a margin 4 cm wide. Manuscript sheets should be numerated consecutively from "1" onwards. Footnotes should be avoided.
5. The titel of the paper should be followed by the author's name (with first name abbreviated), by the institution and its address from which the manuscript originates.
6. Figures and tables should be restricted to the minimum needed to clarify the text. They should be numbered consecutively and must be referred too in the text and on the margin. Figures and tables should be added to the manuscript on separate, consecutively numerated sheets. The tables should have a headline. Legends of figures should be submitted on a separate sheet. All figures should bear the author's name and number of figure overleaf.
   Photographs for half-tone reproduction should be in the form of highly glazed prints.
   Line drawing should be in a form suitable for reproduction. The lettering should be sufficiently large and bold to permit reduction. If requested, original drawing and photographs will be returned to the author upon publication of the paper.
7. Formulae should not be written to small and not with pencil. Separate lines for formulae are desirable. SI-units should be used.
   Letters in formulae are normally printed in italics, numbers in ordinary upright typeface. Underlining to denote special typefaces should be done in accordance with the following code:
   Italics: wavy underlined with pencil (only necessary for symbols in the text)
   Boldface italics (vectors): wavy underlined twice
   Upright letters (all abbreviations like all units (cm, g, ...), all elements and particles (H, He, ..., n, p, ...), elementary mathematical functions like Re, Im, sin, cos, exp, ...): green underlined
   Greek letters: red underlined
   Boldface Greek letters: red underlined twice
   Upright Greek letters (symbols of elementary particles): red and green underlined
   Large letters: underlined with pencil twice
   Small letters: overlined with pencil twice
   (This will be necessary for handwritten letters that do not differ in shape, as *c C*, *k K*, *o O*, *p P*, *s S*, *u U*, *v V*, *w W*, *x X*, *y Y*, *z Z*).
   It will help the printer if the position of subscripts and superscripts is marked with pencil in the following way: $a_{\hat{i}}$, $b^{i}_{\vee}$, $M_{\hat{i}}{}^{j}_{\vee}$, $M_{\hat{ij}}$, $W_{n}{}^{\hat{}}{}_{\hat{i}}$
   Please differentiate between following symbols: $\boldsymbol{a}$, α; $\boldsymbol{\alpha}$, $\propto$, ∞; α, $\boldsymbol{d}$; $\boldsymbol{c}$, $\boldsymbol{C}$, $\subset$; e, l; $\mathscr{E}$, ε, $\in$; $\boldsymbol{k}$, $\boldsymbol{K}$, $\varkappa$; $\varkappa$, x, $\boldsymbol{X}$, χ, $\times$; l, 1; o, O, σ, 0; p, ϱ; $\boldsymbol{u}$, $\boldsymbol{U}$, $\cup$; v, $\boldsymbol{\nu}$, $\vee$; φ, ϕ, Φ, $\emptyset$; ϑ, θ, Θ.
8. Each paper should be followed by a list of references with consecutive numeration. The numbers should be placed on the line between squared brackets or typewritten inclined lines. Articles in journals should be cited with author's name and abbreviated first name, title of journal, volume number (underlined), year of publication (in brackets) and page number. In case of books authors' name and abbreviated first name, title, place, and year of publication should be given.
9. Manuscripts should be submitted fit for printing; badly arranged manuscripts are returned.
10. Proof reading should be limited to the correction of typographical errors. Deletions and insertions are not permitted in proof reading.
11. Of every paper published the author(s) will receive 30 reprints free of charge.
12. Within the framework of legal protection the publishers have the right of publication, distribution and translation reserved. Without express permission it is not allowed to produce photocopies, microfilms etc. of this journal or parts of it.

Fortschr. Phys. 32 (1984) 1, 1—24

Uniqueness and Global Almost Markov Property for Regularized Yukawa Gases

R. GIELERAK and B. ZEGARLIŃSKI

*Institute of Theoretical Physics, University of Wroclaw, Wroclaw, Poland**)

Using a functional description of the Grand Canonical Ensemble we prove the uniqueness of the Dobrushin-Lanford-Ruelle equations for some class of the regularized Yukawa-like gases. The uniqueness result holds in this region of chemical activities and temperatures where the contraction map principle can be applied to the Kirkwood-Salsburg equations. Additionally we prove Almost Global Markov Property for such gases.

1. Introduction

One of the main problems in the statistical physics are the questions about existence and uniqueness of the Gibbs states. In the case of lattice systems there exists a large variety of results concerning these problems [*1, 2, 3, 4, 5*]. However in the case of continual systems the situation is not so clear. One of the main difficulties here is to work with very complicated configuration spaces for such a systems. Especially complicated configuration spaces are used to describe grand canonical Gibbs ensemble for the continual systems [*6, 7*].

Whenever two-body potential is function of positive type there exists a very elegant and compact description of the grand canonical Gibbs ensemble. This description was used for the first time by SIEGERT [*8*]. Using these methods there has been obtained in the last years a rich variety of new results concerning the existence and shape independence of the thermodynamic limit for the pressure and correlation functions [*9, 10, 11*]. These methods have been used to obtain new correlation inequalities for two-component (neutral) classical and quantum mechanical systems of distinguishable particles [*12*]. It is the aim of the present paper to use functional integrals in order to discuss existence and uniqueness of the solutions of the Dobrushin-Lanford-Ruelle equations [*1, 2, 3*] for some continual systems. Inspiration for our investigation comes from the paper by ALBEVERIO and HOEGH-KROHN [*13*] where they proved uniqueness of the (regular) solutions of the Dobrushin-Lanford-Ruelle equations for the two-dimensional Yukawa gas [*10, 11*] in the region of couplings where the Glimm-Jaffe-Spencer expansion converges [*14*]. Let us describe framework in which we shall to work and results we have obtained in this paper.

Let $Q = \mathsf{R}$ and the potential V on $(Q \times \mathsf{R}^d)^2$ be some function of positive type satisfying:

$$V(q, x; q', x') = q \cdot q'(2\pi)^{-\alpha/2} \int e^{ik(x-x')}\hat{V}(k)\, d^dk \tag{1.1}$$

*) 50-205 Wroclaw, Cybulskiego 36

where $0 < V(k) \in L^1(\mathsf{R}^d)$. Let λ be some finite measure on Q. By $\mathscr{S}(\mathsf{R}^d)$ we denote the Schwartz space of infinitely many times differentiable functions on R^d which are rapidly decreasing at ∞. By $\mathscr{S}'(\mathsf{R}^d)$ we mean the dual of $\mathscr{S}(\mathsf{R}^d)$ i.e. the space of tempered distributions. By Minlos theorem the functional

$$J_V(f) = e^{-1/2(f,Vf)}. \tag{1.2}$$

where $f \in \mathscr{S}(\mathsf{R}^d)$,

$$(f, Vf) = \int dx\, dy f(x)\, V(x-y)\, f(y) \tag{1.3}$$

is the characteristic functional of a Gaussian measure $d\mu_V$ with mean 0 and covariance V and with support contained in $\mathscr{S}'(\mathsf{R}^d)$. The expectation with respect to $d\mu_V$ is defined by

$$E_V(F) = \int_{\mathscr{S}'(\mathsf{R}^d)} d\mu_V(\phi)\, F(\phi) \tag{1.4}$$

for $F \in L^1(\mathscr{S}', d\mu_V)$.
Wick ordering is defined by:

$$:e^{i\phi(f)}: = e^{1/2(f,Vf)} e^{i\phi(f)}. \tag{1.5}$$

Let us define also

$$U_\Lambda = z \int_Q d\lambda(q) \int_\Lambda dx\, :e^{i\beta^{1/2} q\phi(x)} \tag{1.6}$$

The grand canonical partition function in a compact region $\Lambda \subset \mathsf{R}^d$ for the system of particles in the thermal equilibrium interacting via two-body potential $qq' V(x, x')$ can be written as:

$$Z_\Lambda(z, \beta) = \int_{\mathscr{S}'} \mathrm{d}\mu_V(\phi) \exp(U_\Lambda) \tag{1.7}$$

and the (particle-) density correlation functions for this system as:

$$\varrho_\Lambda(\beta, z)\big((q)_n, (x)_n\big) = \frac{z^n}{Z_\Lambda(\beta, z)} \int_{\mathscr{S}'(\mathsf{R}^d)} d\mu_V(\phi) \prod_{i=1}^{n} :\mathrm{e}^{i\beta^{1/2} q_i \phi}:(x_i) \exp U_\Lambda(\phi) \tag{1.8}$$

where $(q)_n := (q_1, \ldots, q_n)$ and $(x)_n = (x_1, \ldots, x_n)$.
From this formulas it arises a natural interpretation of $\mathscr{S}'(\mathsf{R}^d)$ as a configuration space for our system in the grand canonical Gibbs ensemble. The factor $\exp U_\Lambda(\phi)$ can be interpreted formally as a Gibbs factor and $d\mu_V(\phi)$ as a density of states for the free system.
To formulate Dobrushin-Lanford-Ruelle equations in this framework we require that V is the covariance of an almost Markov Gaussian field. To justify this concept we introduce some more notations. Let $\Sigma(\Lambda)$ be a local σ-algebra generated by the field ϕ. Let $(\Lambda)_\varepsilon$ means the ε-neighbourhoods of the given set $\Lambda \subset \mathsf{R}^d$ i.e. $(\Lambda)_\varepsilon := \{x \in \mathsf{R}^d : \operatorname{dist}(x, \Lambda) \leq \varepsilon\}$. By $E_V{}^0\{\cdot \mid \Lambda\}$ be denote the conditional expectation values of the measure $d\mu_V$ with respect to the local σ-algebra $\Sigma(\Lambda)$. We say that the Gaussian measure $d\mu_V$ is an almost (ε-) Markovian on $\mathscr{S}'(\mathsf{R}^d)$ iff for any open $\Lambda \subset \mathsf{R}^d$ and F the random variable measurable with respect to $\Sigma(\Lambda)$ we have

$$\begin{gathered} E_V{}^0\{F \mid \Lambda^c\} = E_V{}^0\{F \mid (\partial\Lambda)_\varepsilon{}^+\}, \\ (\partial\Lambda)_\varepsilon{}^+ := \Lambda^c - (\Lambda_\varepsilon)^c. \end{gathered} \tag{1.9}$$

The complete description of all almost Markovian Gaussian measures has been done by MOLCZAN in [*15*]. There exists also a complete description of the Markov-Gauss field

Volume 31 1983

ISSN 0015-8208 Fortschr. Phys., Berlin **31** (1983)

Zeitschrift „Fortschritte der Physik"

Herausgeber: Prof. Dr. Frank Kaschluhn, Prof. Dr. Artur Lösche, Prof. Dr. Robert Rompe; Berlin.
Verlag: Akademie-Verlag, DDR-1086 Berlin, Leipziger Straße 3–4; Fernruf: 2236221 und 2236229; Telex-Nr.: 114420;
Bank: Staatsbank der DDR, Berlin, Konto-Nr.: 6836-26-20712.
Chefredakteur: Dr. Lutz Rothkirch.
Anschrift der Redaktion: Sektion Physik der Humboldt-Universität zu Berlin, DDR-1040 Berlin, Hessische Straße 2.
Veröffentlicht unter der Lizenznummer 1324 des Presseamtes beim Vorsitzenden des Ministerrates der Deutschen Demokratischen Republik.
Gesamtherstellung: VEB Druckhaus „Maxim Gorki", DDR - 7400 Altenburg, Carl-von-Ossietzky-Straße 30/31.
Erscheinungsweise: Die Zeitschrift „Fortschritte der Physik" erscheint monatlich: Die 12 Hefte eines Jahres bilden einen Band. Bezugspreis je Band 216,– M zuzüglich Versandspesen (Preis für die DDR: 120,– M). Preis je Heft 18,– M (Preis für die DDR: 10,– M).
Bestellnummer dieses Bandes: 1027/31.

AN (EDV) 57618

Inhalt des 31. Bandes

Number 1

Number 2

Number 3

Number 4

Number 5

Number 6

Number 7

Number 8—9

Number 10

Number 11

Number 12

on $\mathscr{S}'(\mathsf{R}^d)$ i.e. the almost Markovian Gauss fields with $\varepsilon = 0$. See [*16, 17*]. Let us denote

$$d\mu_\Lambda(\phi) := \frac{1}{Z_\Lambda} \cdot \exp\left(U_\Lambda(\phi)\right) \cdot d\mu_V(\phi). \tag{1.10}$$

Roughly speaking we define now the Gibbs measure corresponding to our system. By this we means every measure $d\mu$ on $\mathscr{S}'(\mathsf{R}^d)$ which fulfill:

1) $\mu_{\upharpoonright\Sigma(\Lambda)}$ is absolutely continuous with respect to $\mu_{V\upharpoonright\Sigma(\Lambda)}$ for Λ bounded.
2) DLR equation

If F is measurable with respect to $\Sigma(\Lambda)$, Λ-bounded, then

$$E_\mu\{F \mid \Lambda^c\} = E_\mu\{F \mid (\partial\Lambda)_\varepsilon^+\} = E_{\mu_\Lambda}\{F \mid (\partial\Lambda)_\varepsilon^+\} \tag{1.11}$$

where $E_\mu\{\cdot \mid \Lambda\}$ means conditional expectation value with respect to measure $d\mu$ and the σ-algebra $\Sigma(\Lambda)$. We say that the Gibbs measure μ is pure iff there is no other measure μ' on $\mathscr{S}'(\mathsf{R}^d)$ fulfilling (1) and (2) and such that μ' is absolutely continuous with respect to μ.
In this paper we consider the problem of uniqueness of the (regular) solutions of the DLR equations for some special choice of V. Namely we consider here regularized Yukawa gases i.e. we consider a systems with two body potential given by

$$V(x, x') = \left(h_\varepsilon * (-\Lambda + m^2)^{-1} * h_\varepsilon\right)(x, x') \tag{1.12}$$

where $*$ means the convolution and h_ε is smooth function with compact support of diameter ε. For more details see Section 3 of this paper. The regularization is necessary in dimension $d > 2$ because otherwise the system is not stable. The limitation of our paper to such systems is of technical nature. In separate paper we hope to present an extension of our results to more general covariances [*18*]. Kirkwood-Salsburg equations are applicable to controll thermodynamic limit for the measures $d\mu_\Lambda(\phi)$. For $\lambda > 0$ one can use also the correlation inequalities founded by Fröhlich and Park [*9*]. In the case of regularized Yukawa-like systems we are able to prove following uniqueness result.

Theorem 1.1. For Yukawa-like systems (1.12) in the region where the contraction map principle can be applied to the Kirkwood-Salsburg equations the obtained Gibbs measure $d\mu_\varepsilon$ is pure.
In the two-dimensional situation this theorem have been proved by Albeverio and Hoegh-Kröhn for systems without any regularization and some restriction on the size of support of the measure $d\boldsymbol{\lambda}(\cdot)$. They used powerful technic of Glimm-Joffe-Spencer cluster expansion. We prove our result by analysis of the Kirkwood-Salsburg operators [*19*]. See Section 5.
Finally we consider the problem of the Global Almost Markov property for the measure $d\mu_\varepsilon$. Global Markov property plays very important role in the field theory [*20, 21*]. As far as we know the only existed proofs of this property in the Euclidean field theory are the one in the case of weakly coupled two-dimensional sine-Gordon model given by Albeverio Hoegh-Kröhn in [*13*] and one in the case of exponential interaction given by the one of authors of the present paper in [*22*]. In the statistical mechanics Global Markov property implies the existence of the transfer matrix [*23*]. Let us justify briefly the concept of the Global Almost Markov Property (GAMP for short hand). Let C be a given smooth hypersurface in R^d which divide R^d into two disjoint open sets Ω^- and Ω^+. We say that the Gibbs measure $d\mu$ has GAMP iff for any F_1 (F_2) measur-

able with respect to $\Sigma(\Omega^+ \setminus (C)_\varepsilon)$ $(\Sigma(\Omega^- \setminus (C)_\varepsilon)$ respectively) we have

$$E_\mu\{F_1F_2 \mid (C)_\varepsilon\} = E_\mu\{F_1 \mid (C)_\varepsilon\} \cdot E_\mu\{F_2 \mid (C)_\varepsilon\}. \tag{1.13}$$

In the section 6 of this paper we prove our second main result namely the following theorem.

Theorem 1.2. The infinite volume measure $d\mu_\varepsilon$ for Yukawa-like systems (1.12) in the region of chemical activities where the contraction map principle can be applied to the Kirkwood Salsburg equations has GAMP.

2. The Almost Markovian Gibbs Measures

Let ϕ be a Markov field [25, 26] over $\mathscr{S}(\mathrm{R}^d)$ on the probability space $\{Q, \Sigma, d\mu_0\}$. Let $\Sigma(\Lambda)$, where $\Lambda \subset \mathrm{R}^d$ is an open set, be a σ-algebra generated by the random variables $\phi(f)$ with supp $f \subset \Lambda$. For closed $\Lambda \subset \mathrm{R}^d$ we define $\Sigma(\Lambda)$ as an intersection of $\Sigma(\Lambda)'$ where $\Lambda' \subset \mathrm{R}^d$ runs over all open sets containing Λ in their interiors. If the random variable F is measurable with respect to the σ-algebra $\Sigma(\Lambda)$ we write $F \in m(\Sigma(\Lambda))$.

Definition 2.1. A random variable V is an almost (ε-) additive functional of the field ϕ iff for each finite cover $\{\Lambda_i\}$ of the space R^d there exist random variables $U_i \in m(\Sigma((\Lambda_i)_\varepsilon))$, where $(\Lambda)_\varepsilon = \{x \in \mathrm{R}^d : \text{dist}\,(x, \Lambda) \leqq \varepsilon\}$ such that

$$V = \sum_i U_i.$$

In the following we consider these almost additive functionals which obey $\exp(-\lambda V) \in L(d^1\mu_0)$ for some $\lambda \in \mathrm{R}(C)$. Having such an almost additive functional V of the field ϕ we define a new measure on $\{Q, \Sigma\}$ by the formula

$$\begin{aligned} d\mu_V(\phi) &= \frac{1}{Z} e^{-\lambda V(\phi)} \cdot d\mu_0(\phi) \\ Z &= \int_{\mathscr{S}'(\mathrm{R}^d)} \exp(-\lambda V) \cdot d\mu_0(\phi). \end{aligned} \tag{2.1}$$

For $F \in L^1(d\mu)$ and $\Lambda \subset \mathrm{R}^d$ we denote by $E_V\{F \mid \Lambda\}$ $(E_0\{F \mid \Lambda\})$ the conditional expectation values with respect to the measure μ_V and the σ-algebra $\Sigma(\Lambda)$ (μ_0 and $\Sigma(\Lambda)$ respectively).

Proposition 2.1. Let ϕ be a Markov random field over $\mathscr{S}(\mathrm{R}^d)$ given on the probabilistic space $\{Q, \Sigma, \mu_0\}$. Let μ_V be a probabilistic measure given by an almost additive functional V.
Then the random variable ϕ given on $\{Q, \Sigma, \mu_V\}$ has an Almost (ε-) Markov Property (i.e. for all $F \in L^1(d\mu)$ and $\Lambda \subset \mathrm{R}^d$ we have:

$$\begin{aligned} &E_V\{E_V\{F \mid \Lambda\} \mid (\Lambda^c)_\varepsilon\} \\ &= E_V\{E_V\{F \mid \Lambda^c\} \mid \Lambda_\varepsilon\} \\ &= E_V\{F \mid (\partial\Lambda)_\varepsilon\}. \end{aligned} \tag{2.2}$$

Proof:
The proof is standard (see e.g. [2, 6]) but for completeness let us recall of.
Let $\alpha \in m(\Sigma(\Lambda))$ and let α be a bounded. By the Radon-Nikodym theorem there exists one and only one random variable $\tilde{\alpha} = E\{\alpha \mid (\Lambda^c)_\varepsilon\}$ such that for each $\gamma \in m(\Sigma((\Lambda^c)_\varepsilon))$

and γ bounded we have:

$$E_V \gamma \alpha = E_V \gamma \tilde{\alpha} \tag{2.3}$$

which is equivalent to:

$$\frac{1}{Z} \int_Q \gamma \alpha e^{-\lambda V} d\mu_0(\phi) = \frac{1}{Z} \int_Q \gamma \tilde{\alpha} e^{-\lambda V} d\mu_0(\phi). \tag{2.4}$$

Applying again the Radon-Nikodym theorem we get:

$$\int_Q d\mu_0(\phi)\, \gamma \alpha e^{-\lambda V} = \int_Q d\mu_0(\phi)\, \gamma E_0\{\alpha e^{-\lambda V} \mid (\Lambda^c)_\varepsilon\} \tag{2.5}$$

and

$$\int_Q d\mu_0(\phi)\, \gamma \tilde{\alpha} e^{-\lambda V} = \int_Q d\mu_0(\phi)\, \gamma \tilde{\alpha} E_0\{e^{-\lambda V} \mid (\Lambda^c)_\varepsilon\}. \tag{2.6}$$

Because $e^{-\lambda V}$ is strictly positive it yields:

$$E_V\{\alpha \mid (\Lambda^c)_\varepsilon\} = \frac{E_0\{\alpha e^{-\lambda V} \mid (\Lambda^c)_\varepsilon\}}{E_0\{e^{-\lambda V} \mid (\Lambda^c)_\varepsilon\}}. \tag{2.7}$$

Now we prove that the r.h.s. of (2.7) is in $m\big(\Sigma((\partial\Lambda)_\varepsilon)\big)$. Let Λ_1 and Λ_2 be open sets such that $\Lambda_1 \cup \Lambda_2 = \mathbf{R}^d$, $\bar{\Lambda} \subset \Lambda_1$ and $\Lambda_2 \subset (\Lambda_\varepsilon)^c$ where $\bar{\Lambda}$ is the closure of Λ. Then there exist Λ_1 and V_2 such that $V = V_1 + V_2$ and $V_1 \in m\big(\Sigma(\Lambda\ )_\varepsilon\big)$, $V_2 \in \overline{m}\big(\Sigma(\Lambda_2)_\varepsilon\big) \subset m\big(\Sigma(\Lambda^c)_\varepsilon\big)$. From the projection property of the conditional expectation values it follows

$$E_V\{\alpha \mid (\Lambda^c)_\varepsilon\} = \frac{E_0\{\alpha e^{-\lambda V_1} \mid (\Lambda^c)_\varepsilon\}}{E_0\{e^{-\lambda V_1} \mid (\Lambda^c)_\varepsilon\}}. \tag{2.8}$$

By the Markov property of the measure μ_0 and Lemma 2.2 below $E_V\{\alpha \mid (\Lambda^c)_\varepsilon\}$ is in $\overline{m}\big(\Sigma(\Lambda^c)_\varepsilon \cap \Lambda_1\big)$ for all Λ_1 as described above. Hence $E_V\{\alpha \mid (\Lambda^c)_\varepsilon\}$ is in $m\big(\Sigma(\partial\Lambda)_\varepsilon\big)\}$. Now if $\alpha = E_V\{F \mid \Lambda\}$ for some $F \in L^1(d\mu_\nu)$ then

$$E_V\big\{E_V\{F \mid \Lambda\} \mid (\Lambda^c)_\varepsilon\big\} = E_V\{F \mid (\partial\Lambda)_\varepsilon\}. \tag{2.9}$$

The second equality in (2.2) can be proved by analogous arguments.

Lemma 2.2

Let the measure μ_0 on $\{Q, \Sigma\}$ has MP i.e. for all $\alpha \in m\big(\Sigma(\Lambda)\big)$ (Λ — open in $\mathbf{R}^d$) and $\alpha \in L^1(d\mu_0)$ $E_0\{\alpha \mid \Lambda\} = E_0\{\alpha \mid \partial\Lambda\}$ then for $\alpha \in m\big(\Sigma(\Lambda)_\varepsilon\big)$, Λ-open

$$E_0\{\alpha \mid (\Lambda^c)_\varepsilon\} = E_0\{\alpha \mid (\partial\Lambda)_\varepsilon\}. \tag{2.10}$$

Proof:

For the standard proof we refer to [*26*], proofs of *Lemmas* 1 and 2.

As we see below one can define a compatible set of transition functions with the aid of the almost-additive functional. This enables us to define the set of Gibbs measures characterized completely by these transition functions.

Definition 2.2

1. Let ϕ be a Markov field over $\mathcal{S}(\mathbf{R}^d)$ given on $\{Q, \Sigma, \mu_0\}$. For any regular $\Lambda \subset \mathbf{R}^d$ and for (almost) each $\eta \in \Omega$ we define a new measure $d\mu_0{}^\Lambda(\cdot \mid \eta)$ by the generating functional

$$\int e^{i\phi(f)} d\mu_0{}^\Lambda(\phi) \mid \eta) := E_0\{e^{i\phi(f)} \mid \Lambda\}\,(\eta). \tag{2.11}$$

2. Let V be an almost additive functional of the Markov field ϕ over $\mathcal{S}(\mathsf{R}^d)$ given on $\{Q, \Sigma, d\mu_0\}$. We define a measure $d\mu^\Lambda(\cdot \mid \eta)$ by:

$$d\mu^\Lambda(\cdot \mid \eta) = \frac{e^{-\lambda V}}{E_0\{e^{-\lambda V} \mid \Lambda\}\,(\eta)}\, d\mu_0{}^\Lambda(\cdot \mid \eta)\,. \tag{2.12}$$

Proposition 2.4.

For almost all (with respect to $d\mu_0$) η and for all regular $\Lambda \subset \mathsf{R}^d$ the measure $d\mu_0{}^\Lambda(\cdot \mid \eta)$ has Local Almost Markov Property (LAMP).

Proof:

An AMP of the measure $d\mu^\Lambda(\cdot \mid \eta)$ follows from the MP of $\mu_0(\cdot \mid \eta)$ and *Proposition 2.1.* The MP of $\mu_0(\cdot \mid \eta)$ follows from the MP of μ_0 and definition (2.12).

Definition 2.3.

For any open, bounded $\Lambda_0 \subset \mathsf{R}^d$ we define $P_{\Lambda_0}(\Lambda)$ as a set of all probability measures on $\Sigma(\Lambda_0)$ which are of the form

$$\int \mu^{\Lambda^c}(\xi \mid \eta)\, \varrho(\eta)\, d\mu_0(\eta) \tag{2.13}$$

where $\Lambda \supset \Lambda_0$, and $\mu^{\Lambda^c}(\cdot \mid \eta)$ is given by (2.12), $\varrho(\cdot)$ is an arbitrary positive normalized function from $L_1(d\mu_0)$.

Proposition 2.5.

The elements of $P_{\Lambda_0}(\Lambda)$ restricted to $\Sigma(\Lambda_0)$ are absolutely continuous with respect to $\mu_{0|\Sigma(\Lambda_0)}$ and have LAMP i.e. for all $(\Lambda_1)_\varepsilon \subset \Lambda_0$ and $F \in m(\Sigma(\Lambda_1))$

$$E'\{F \mid (\Lambda_1{}^c)_\varepsilon\} = E'\{F \mid (\partial\Lambda_1)_\varepsilon\} \tag{2.14}$$

where $E'\{\cdot \mid \cdot\}$ stand for the conditional expectation values with respect to measures from $P_{\Lambda_0}(\Lambda)$.

Proof:

The LAMP follows form that of $d\mu^{\Lambda^c}(\cdot \mid \eta)$. Because $d\mu_0(\cdot \mid \Lambda)$ restricted to $\Sigma(\Lambda_0)$ are absolutely continuous with respect to $\mu_0 \mid \Sigma(\Lambda_0)$ it follows that element of $P_{\Lambda_0}(\Lambda)$ also are.

It follows from the *Proposition 2.3* that $P_{\Lambda_0}(\Lambda)$ is a decreasing function of Λ. Let us define

$$P_{\Lambda_0} := \bigcap_{\Lambda} P_{\Lambda_0}(\Lambda)\,. \tag{2.15}$$

Proposition 2.6

If $\Lambda_0 \subset \Lambda_1$ then the elements of P_{Λ_0} are the restriction to $\Sigma(\Lambda_0)$ of the elements from P_{Λ_1}.

Proof:

It follows immediately from the definition (2.15).

Definition 2.4

By the Gibbs measures given by an almost additive functional V and the measure μ_0 on $\{Q, \Sigma, \mu\}$ we mean all the measures on $\{Q, \Sigma, \mu\}$ such that their restriction to $\Sigma(\Lambda_0)$ are contained in P_{Λ_0}. The set of all Gibbs measures given by V and μ_0 we denote by $\mathcal{G}\{V, \mu_0\}$. We say that $\mu \in S(V, \mu_0)$ is a pure phase if there is no other elements in $\mathcal{G}$ which are absolutely continuous with respect to μ.

Proposition 2.7

$\mu \in \mathcal{G}(V, \mu_0)$ is a pure phase iff

$$E_\mu\{\cdot \mid \Lambda^c\} \to E_\mu(\cdot) \tag{2.16}$$

as $\Lambda \nearrow \mathbb{R}^d$, where Λ runs through any filter of bounded open sets of $\mathbb{R}^d$.

Proof:

Let $\mu \in \mathcal{G}\{V, \mu_0\}$ and let v be absolutely continuous with respect to μ. If $\Lambda \subset \mathbb{R}^d$ is open and bounded then by definition $v_{|\Sigma(\Lambda)} \sim \mu_{|\Sigma(\Lambda)}$ which means that there exists a function $\varrho \in (\Sigma(\Lambda^c) \cup J_\mu)$ (where J_μ means the ring of $\mu - 0$ measure sets in Σ) such that:

$$dv = \varrho d\mu .$$

Hence

$$\varrho \in \Sigma_\mu{}^\infty = \bigcap_\Lambda \Sigma(\Lambda^c) \cup J_\mu . \tag{2.17}$$

This means that μ is a pure phase iff $\mu_{|\Sigma_\mu{}^\infty}$ is the trivial probability measure i.e. the only $\Sigma_\infty{}^\mu$ measurable functions are constants. Because

$$E_\mu\{\cdot \mid \Lambda^c\} \xrightarrow[\Lambda \uparrow \mathbb{R}^d]{} E_\mu\{\cdot \mid \Sigma_\infty{}^\mu\}$$

monotonously as projection in $L^2(d\mu)$ we conclude that $\mu_{|\Sigma_\mu{}^\infty}$ is trivial iff

$$E_\mu\{\cdot \mid \Lambda^c\} \to E_\mu(\cdot)$$

q.e.d.

3. Gibbs Measures for Classical Gases in the Grand Canonical Gibbs Ensemble

We assume that V is some function of positive type, i.e. for every sequence of complex numbers $\{c_i\}_{i=1}^N$ and sequence of points $\{x_i\} \subset \mathbb{R}^d$ we have

$$\sum_{i,j} c_i \bar{c}_j V(x_i - x_j) \geqq 0 . \tag{3.1}$$

We assume also that the function V depends only on the difference $x - y$ and $\hat{V}(k) \in L^1(\mathbb{R}^d)$.

By Minlos theorem the functional

$$J_V{}^0(f) = \exp\left[-\frac{1}{2}(f, Vf)\right] \tag{3.2}$$

on $\mathcal{S}(\mathbb{R}^d)$ is the characteristic functional of a Gaussian measure $d\mu_V{}^0$ with mean zero and covariance V. Let us consider the following additive functional

$$U_\Lambda = \lambda \int dr(q) \int_\Lambda :e^{iq\phi}: (x)\, dx \tag{3.3}$$

of the Gauss field ϕ corresponding to the measure $d\mu_V{}^0$. We shall assume that dr is some bounded, complex measure on $\mathbb{R}^1$ with compact support on $\mathbb{R}^1$ and such that $\overline{dr(q)} = d\lambda(-q)$.

The perturbed measure

$$d\mu_\Lambda(\phi) = \frac{1}{Z_\Lambda}\left(\exp U_\Lambda(\phi)\right) d\mu_V{}^0(\phi) \tag{3.4}$$

where

$$Z_\Lambda = \int d\mu_V{}^0(\phi) \exp\left(U_\Lambda(\phi)\right) \tag{3.4a}$$

is well defined for all $\Lambda \subset \mathbb{R}^d$ bounded. Making out the shift transformation

$$\phi(x) \to \phi(x) - i(V * f)(x) \tag{3.5}$$

and calculating the corresponding Radon-Nikodym derivative we find the following formula for the characteristic functional of the perturbed measure $d\mu_{V,\Lambda}$:

$$J_\Lambda(f) = e^{-1/2(f,Vf)} \cdot \int d\mu_\Lambda(\phi) \exp\left\{\lambda \int dr(q) \int_\Lambda dx(e^{-q(f*V)(x)} - 1) : e^{iq\phi(x)}:\right\} \tag{3.6}$$

from which after the expansion of the exponential into the powers of λ we get:

$$J_\Lambda(f) = e^{-1/2(f,Vf)} \cdot \sum_{n=0}^{\infty} \frac{1}{n!} \int dr(q)_n \int_\Lambda d(x)_n \prod_{i=1}^{n} [e^{-q_i(f*V)(x_i)} - 1] \cdot \varrho_\Lambda\left((x)_n, (q)_n\right) \tag{3.7}$$

where

$$\varrho_\Lambda\left((x)_n, (q)_n\right) = \lambda^n \int d\mu_\Lambda(\phi) \prod_{i=1}^{n} : e^{iq_i\phi(x_i)}: \tag{3.8}$$

are the (particle-) density correlation functions of a gas of classical particles interacting via twobody interaction of the form

$$U = \sum_{i \neq j} q_i q_j V(x_i - x_j). \tag{3.9}$$

The q's correspond to an internal freedom, for each particle, and the measure $dr(q)$ gives the range of variation for this internal degree of freedom.
Applying again the shift transformation

$$\phi(x) \to \phi(x) - iV(x - x_j) \tag{3.10}$$

in the formula defining $\varrho_\Lambda\left((x)_n, (q)_n\right)$ we have the following equalities

$$\begin{aligned} \varrho_\Lambda\left((x)_n, (q)_n\right) &= \lambda \exp q_1 \sum_{i=2}^{n} q_i V(x_1 - x_i) \cdot \int_{\mathcal{S}'(\mathbb{R}^d)} \prod_{i=2}^{R} : e^{iq_i\phi(x_i)}: \\ &\times \exp\left\{\lambda \int dr(q) \int_\Lambda dx \cdot (e^{q_1 qV(x-x_1)} - 1) : e^{iq\phi(x)}:\right\} \cdot d\mu_\Lambda(\phi) \end{aligned} \tag{3.11}$$

in which after expansion of the exponential into the powers of λ we easily recognize the well known Kirkwood-Salsburg identities. Because of assumption $\hat{V}(\cdot) \in L^1(\mathbb{R}^d)$ the stability condition holds. Therefore contraction map'principle can be applied in order to controll thermodynamic limit for the correlation functions ϱ_Λ. More precisely we get that if $r(q)$ has support in the interval $[-r, r]$ and

$$C_V = \sup_\alpha \int |e^{-\alpha q V(x)} - 1| \, d\,|r|\,(q)\, dx \tag{3.12}$$

is finite, then the correlation function $\varrho_\Lambda\left((x)_n, (q)_n\right)$ converges uniformly on compact subsets to a limit $\varrho\left((x)_n, (q)_n\right)$ as $\Lambda \nearrow \mathbb{R}^d$, if $|\lambda| < C_\Lambda^{-1} \exp\left(-2r^2 V(0) - 1\right)$. Moreover the limit $\varrho\left((x)_n, (q)_n\right)$ is translation invariant in x and analytic in λ for λ in the complex disc $|\lambda| < C_V^{-1} \exp\left(-2r^2 V(0) - 1\right)$. This enables us to prove the following Lemma:

Lemma 3.1

1. For $|\lambda| < C_\Lambda^{-1} \exp\left(-2r^2 V(0) - 1\right)$ the unique limit $\lim d\mu_{V',\Lambda} = d\mu_V$ as weak limit exists. The characteristic functional of the measure $d\mu_V$ is given by:

$$J_V(f) = \exp\left[-\frac{1}{2}(f, Vf)\right] \cdot \sum_{n=0}^{\infty} \frac{1}{n!} \int d\lambda(q)_n \int d(x)_n \prod_{i=1}^{n} [e^{-q_i(V*f)(x_i)} - 1] \cdot \varrho\big((x)_n, (q)_n\big). \tag{3.13}$$

2. The measure $d\mu_V$ is translation invariant and is concentrated on the continuous functions.

3. The measure $d\mu_V$ has the cluster property

$$d\mu_V(A, B_a) \to \mu_V(A)\, \mu_V(B)$$

as $a \to \infty$ in R^d, for any measurable sets A, B where B_a is the set B translated by $a \in \mathsf{R}^d$: $B_a = \{f \mid f(x) = g(x - a) \text{ for some } g \in B\}$.

Remark 1.

One can extend the part 1 of the previous lemma to any $\lambda > 0$ by using the correlation inequalities of the Fröhlich-Park type [9].
Assuming additionally that

$$\Gamma(x) = -(2\pi)^{-d/2} \int e^{ikx} \hat{V}(k)^{-1}\, d^d k \geqq 0 \tag{3.14}$$

for $x \neq 0$ (in the distribution sense) one can conclude that the measure $d\mu_V$ has the cluster property for any $\lambda > 0$. This follows from the facts the two-point moment of the measure $d\mu_V$ has cluster property. Assumption (3.14) means that the measure $d\mu_V$ satisfies the FKG correlation inequalities. Therefore applying Lebowitz-Simon theorem [33] we conclude that $d\mu_V(\cdot)$ has cluster, property for any $\lambda > 0$.
It is possible also to relax the assumption that $dr(\cdot)$ has compact support. To cope with this case it suffices to modify the definition of the Banach space used to the analysis of the Kirkwood-Salsburg operators [32].

Remark 2.

For $dr(\cdot) \geqq 0$ one can relate ϱ_Λ, ϱ and the measures $d\mu_{V,\Lambda}$ $d\mu_V$ to the well known quantities introduced in classical statistical mechanics for the description of states [7]. In particular the ϱ determine a Gibbsian translational invariant probability measure dG_V on the space of the denumerable subsets of R^d which have the property that their intersection with any bounded measurable set is finite. dG_V is the limit of $dG_{V,\Lambda}$, in the sense that $G_{V,\Lambda}(A) \to G(A)$, uniformly in A, for any bounded Lebesque measurable set A in R^d, where $dG_{\Lambda,V}$ is the Gibbsian measure determined by ϱ_Λ:

$$\int_A dG_{V,\Lambda} = \sum_{k=0}^{|A|} \frac{\lambda^k}{k!} \int_\Lambda dr(q)_k \int_\Lambda d(x)_k \cdot \sum_{p=0}^{\infty} (-1)^p \int dr(q)_k \int_\Lambda d(x')_p\, \varrho_{V,\Lambda}\big((x)_k, (x')_p, (q)_k, (q')_p\big) \tag{3.15}$$

and $|A|$ is the number of points in A.
Let $\mathscr{P}(\partial/\partial x_i)$ be an positive elliptic differential operator in $\mathscr{L}^2(\mathsf{R}^d)$ which is translation invariant. Then as it follows from [16] (see also [17]) the corresponding Gaussian process on $\mathscr{S}(\mathsf{R}^d)$ with covariance

$$V(x - x') = \frac{1}{(2\pi)^{d/2}} \int \frac{e^{ik(x-x')}}{\mathscr{P}(..k_i)}\, d^d k$$

has Markov Property. Assuming $V(0) < \infty$ we can apply the discussion of the previous section with $\varepsilon = 0$. This yields to the following result.

Proposition 3.2.
The infinite volume measure $d\mu_V$ corresponding to the interaction (3.16) is Gibbs measure which has Local Markov Property.
The question whether the Gibbs measures from this proposition have Global Markov property arises. However this question is closely connected with the question whether these Gibbs measures are pure. We have no general answer on this question.
In the case $V(0) = \infty$ we have to deal with instable system. In order to avoid this difficulty we perform some regularization. Let h_ε be a smooth function with support of size $2 \cdot \varepsilon$. Then because $\mathscr{P}(..k_i..)^{-1}$ is an multiplicator on $\mathscr{S}(\mathsf{R}^d)$ we have that $|\hat{h}_\varepsilon|^2/\mathscr{P}(k_i) \in \mathscr{S}(\mathsf{R}^d)$ which implies that

$$V_\varepsilon(x, x') = \frac{1}{(2\pi)^{d/2}} \int \frac{|h_\varepsilon(k)|^2}{\mathscr{P}(k_i)} e^{ik(x-x')} d^dk = (h_\varepsilon * V * h_\varepsilon)(x, x') \tag{3.16}$$

is stable potential. From the other hand this kind of regularization implies that V_ε defines a covariance of an Almost Markov Gauss field on $\mathscr{S}(\mathsf{R}^d)$. See [*15*].
Let $\phi_\varepsilon(x) = (\phi * h_\varepsilon)(x)$. Then we have (for $F \in L^1(d\mu_V)$)

$$\int_{\mathscr{S}'(\mathsf{R}^d)} d\mu_V(\phi)\, F(\phi_\varepsilon) \exp \lambda \int dr(q) \int_\Lambda :e^{iq\phi_\varepsilon(x)}: dx$$
$$= \int_{\mathscr{S}'(\mathsf{R}^d)} d\mu_{V_\varepsilon}(\bar{\phi})\, F(\phi) \exp \lambda \int dr(q) \int_\Lambda :e^{iq\phi(x)}: dx$$

Therefore we can again apply the discussion of the section 2.

Proposition 3.3
Let two-body potential be of the form $qq'V_\varepsilon(x, x')$ where V is given by (3.16). For $|\lambda| < C_{V_\varepsilon} \exp\left(-2r^2 V_\varepsilon(0) - 1\right)$ the corresponding infinite volume measure $d\mu_{V_\varepsilon}$ has LAMP.
In the following of this paper we concentrate ourselves on the particular choice of V, namely we consider

$$V(x) = \frac{1}{(2\pi)^{d/2}} \int \frac{e^{ikx}}{k^2 + m^2} d^dk \tag{3.17}$$

and we perform regularization described above. The corresponding to this covariance Gauss measure we denote $d\mu_0$ and sometimes we will call them the free field measure. The constructed infinite volume limit in the *Lemma 3.1* we denote by $d\mu_\varepsilon$ and the corresponding characteristic functional by J_ε.

4. Conditional Expectation Values

In this section we describe some more explicite formulas for the conditional expectation values with respect to the free measure μ_0 (see [*13*], [*27*]).
Let C be a smooth hypersurface which divides R^d into two disjoint sets Ω and Ω^C. For any $g \in C_0(\mathsf{R}^d)$ (i.e. the continuous function with compact support) there exists only one [*28*] $\Psi_g{}^C(x)$ such that

$$\begin{cases} (-\Delta + m^2)\, \Psi_g{}^C(x) = 0 & \text{for } x \in \mathsf{R}^2 \setminus C \\ \Psi_g{}^C(x) = g(x) & \text{for } x \in C. \end{cases} \tag{4.1}$$

Moreover there exists $\mathsf{R}^d \times \mathscr{G}$ measurable function (x, z) such that

$$\Psi_g{}^C(x) = \int_C P_C(x, z)\, g(z)\, dz. \tag{4.2}$$

The functional (4.2) can be continued from $C_0(\mathsf{R}^d)$ to the support of the free field measure μ_0 as a continuous, positive linear functional. It follows by an explicite calculation that $\Psi_g{}^C$ is a gaussian field with covariance

$$\delta S^C(x, y) = S(x, y) - S^C(x, y) \tag{4.3}$$

where $S(x, y)$ is the covariance of the free field and $S^C(x, y) = (-\Delta_C + m^2)$ where Δ_C is the ordinary Laplacean with Dirichlet boundary condition on C. The Dirichlet free field with covariance S^C will be denoted by $\phi^C(x)$. One can prove that $\{\psi_\phi{}^C(x), \phi^C(x)\}$ is an orthogonal decomposition of the free field $\phi(x)$ with respect to the σ-algebra $\Sigma(C)$ i.e. $\psi_\phi{}^C$, ϕ^C are stochastically independent and $\phi(x) = \psi_\phi{}^C(x) + \phi^C(x)$. With the aid of this orthogonal decomposition we can write a formula for the conditional expectation values with respect to $\Sigma(C)$:

$$E_0\{F(\phi) \mid C\}\,(\eta) = E_0{}^C\big(F(\phi^C + \psi_\eta{}^C)\big) \tag{4.4}$$

where $E_0{}^C$ is the expectation for the Dirichlet field ϕ^C. For our purposes we need some simple generalization of (4.4.).

Definition 4.1.

$$\psi_\phi{}^{C_\varepsilon}(x) \equiv \begin{cases} \psi_\varphi{}^{\partial(\Omega/C_\varepsilon)}(x) & \text{for } x \in \Omega/C_\varepsilon \\ \phi(x) & \text{for } x \in C_\varepsilon \\ \psi_\varphi{}^{\partial(\Omega_\varepsilon C)}(x) & \text{for } x \in \Omega_\varepsilon{}^C \end{cases} \tag{4.5}$$

$$\phi^{C_\varepsilon}(x) \equiv \begin{cases} \phi^{\partial(\Omega/C_\varepsilon)}(x) & \text{for } x \in \Omega/C_\varepsilon \\ 0 & \text{for } x \in C_\varepsilon \\ \phi^{((\partial\Omega)_\varepsilon)}(x) & \text{for } x \in (\Omega_\varepsilon)^C. \end{cases} \tag{4.6}$$

By straightforward calculation we have

$$\delta S^{C_\varepsilon}(x, y) = E_0\big(\psi_\phi{}^{C_\varepsilon}(x)\, \psi_\phi{}^{C_\varepsilon}(y)\big) = S(x - y) - S^{C_\varepsilon}(x, y) \tag{4.7}$$

where

$$S^{C_\varepsilon}(x, y) \neq E_0\big(\phi^{C_\varepsilon}(x)\, \phi^{C_\varepsilon}(y)\big) = \begin{cases} S^{\partial(\Omega/C_\varepsilon)}(x, y) & \text{for } x, y \in \Omega/C_\varepsilon \\ 0 & \text{for } x \in C_\varepsilon,\, y \notin C_\varepsilon \\ & \text{or } x \notin C_\varepsilon,\, y \in C_\varepsilon \\ S^{(\partial\Omega)_\varepsilon}(x, y) & \text{for } x, y \in \Omega^C/C_\varepsilon. \end{cases} \tag{4.8}$$

It implies the validity of:

Lemma 4.1

$\{\psi^{C_\varepsilon}(x), \phi^{C_\varepsilon}(x)\}$ is an orthogonal decomposition of the free field $\phi(x)$ with respect to the σ-algebra $\Sigma(C_\varepsilon)$.
For each $F \in L^1(d\mu_0)$ and almost every (w.r to $d\mu_0$) $\eta \in \mathscr{S}'(\mathsf{R}^d)$ we have:

$$E_0\{F(\phi) \mid C_\varepsilon\}\,(\eta) = E_0{}^{C_\varepsilon}\big(F(\phi^{C_\varepsilon}(x) + \psi_\phi{}^{C_\varepsilon})\big) \tag{4.9}$$

where $E_0{}^{C_\varepsilon}$ is the mean value with respect to the field ϕ^{C_ε}.

Let now ϕ be a regular random field over $\mathscr{S}(\mathsf{R}^d)$ on $\{\mathscr{S}(\mathsf{R}^d), \Sigma, d\mu\}$ i.e. a field for which we have

$$E_\mu(\phi(f)\,\phi(f)) \leqq a\,\|f\|_{-1}^2 \tag{4.10}$$

for some constant a.
For such field ϕ we define a linear functional $\psi_\phi{}^{C_\varepsilon}$ by the formula (4.5). From the definition of the regularity (4.10) and properties of $\delta S^{C_\varepsilon}(x, y)$ it follows (see [13]).

Lemma 4.2

Let $\{C_n\}$ be a sequence of hypersurfaces as above and let $\{\Lambda_n\}$ be a sequence of bounded sets in R^d with smooth boundaries such that

$$\lim_n \operatorname{dist}(0, C_n) = \infty$$

$$\lim_n \operatorname{dist}(\Lambda_n, C_n) = \infty$$

and $\Lambda_n \nearrow \mathsf{R}^d$ monotonously by inclusion. If ϕ is a regular random field then there exists a subsequence $\{n'\} \subset \{n\}$ such that

$$\lim_{n'\to\infty} \sup_{x\in\Lambda_{n'}} |\psi_\eta{}^{C_{n'}}(x)| = 0 \tag{4.11}$$

for μ-almost all $\eta \in \mathscr{S}'(\mathsf{R}^d)$.
Let C^0 be an unbounded smooth hypersurface which divides R^d into two disjoint sets Ω and Ω^C. Let C be a bounded, smooth hypersurface which divides R^d into two disjoint sets Λ and Λ^C.
We define the following linear functional of the given above regular field ϕ by:

$$\psi_\phi{}^{(C^0,C)_\varepsilon}(x) = \begin{cases} \psi_\phi{}^{\partial(\Lambda\cap\Omega/(C_\varepsilon{}^0\cup C_\varepsilon))}(x) & \text{for } x \in \Lambda\cap\Omega/(C_\varepsilon{}^0\cup C_\varepsilon) \\ \psi_\phi{}^{\partial(\Lambda\cap\Omega^C/(C_\varepsilon{}^0\cup C_\varepsilon))}(x) & \text{for } x \in \Lambda\cap\Omega^C/(C_\varepsilon{}^0\cup C_\varepsilon) \\ \phi(x) & \text{for } x \in C_\varepsilon{}^0\cup C_\varepsilon \\ \psi_\phi{}^{\partial(\Lambda^C\cap\Omega/(C_\varepsilon{}^0\cup C_\varepsilon))}(x) & \text{for } x \in \Lambda^C\cap\Omega/(C_\varepsilon{}^0\cup C_\varepsilon) \\ \psi_\phi{}^{\partial(\Lambda^C\cap\Omega^C/(C_\varepsilon{}^0\cup C_\varepsilon))}(x) & \text{for } x \in \Lambda^C\cap\Omega^C/C_\varepsilon{}^0\cup C_\varepsilon\,. \end{cases} \tag{4.12}$$

If ϕ is a free field we also define

$$\phi^{C_\varepsilon\cup C_\varepsilon}(x) = \begin{cases} 0 \quad \text{for } x \in C_\varepsilon{}^0\cup C_\varepsilon \\ \text{the free field with the Dirichlet boundary} \\ \text{condition on } \partial(C_\varepsilon{}^0\cup C_\varepsilon) \quad \text{for } x \notin C_\varepsilon{}^0\cup C_\varepsilon \end{cases} \tag{4.13}$$

Then for all $F(\phi) \in L^1(d\mu_0)$ we find the following formula for the conditional expectation with respect to $\Sigma(C_\varepsilon{}^0\cup C_\varepsilon)$:

$$E_0\{F(\phi) \mid C_\varepsilon{}^0\cup C_\varepsilon\}(\eta) = E_0{}^{C_\varepsilon{}^0\cup C_\varepsilon}\big(F(\psi_\eta{}^{(C^0\cup C)_\varepsilon}(x) + \phi^{C_\varepsilon{}^0\cup C_\varepsilon})\big) \tag{4.14}$$

where $E_0{}^{C_\varepsilon{}^0\cup C_\varepsilon}$ is the mean value with respect to the Gaussian field $\phi^{C_\varepsilon{}^0\cup C_\varepsilon}$.
Let us denote by $S_0{}^{C_\varepsilon{}^0\cup C_\varepsilon}(x, y)$ the covariance of the field $\Psi^{(C^0,C)_\varepsilon}$. The covariance of the field $\Psi_\phi{}^{(C^0,C)_\varepsilon}(x)$ is equal to $S(x-y) - S^{C_\varepsilon{}^0\cup C_\varepsilon}(x, y)$. For the proof of Global Almost Markov Property we shall need the following Lemma.

Lemma 4.3

Let ϕ be a regular random field on $\{S'(\mathsf{R}^d), \Sigma, d\mu\}$. Let C^0 be a fixed unbounded smooth hypersurface which divides R^d into two disjoint set Ω and Ω^C. Let $\{C_n\}$ and $\{\Lambda_n\}$ be

a sequences as in the Lemma 4.2, and additionally we assume that $C_n \cap C^0$ is $(d-2)$-dimensional.
Then there exists a subsequence $\{n'\} \subset \{n\}$ such that

$$\lim_{n'\to\infty} \sup_{x\in\Lambda_{n'}} |\psi_\eta^{(C^0, C_{n'})_\varepsilon}(x) - \psi_\eta^{C_\varepsilon^0}(x)| = 0 \tag{4.15}$$

for μ-almost all $\eta \in S'(\mathbb{R}^d)$.

5. Uniqueness

In this section we prove one of our main results of this paper. We prove that the measure μ_ε constructed in the section 3 is pure Gibbs measure corresponding to the almost additive functional

$$V_\Lambda(\phi_\varepsilon) = \lambda \int_\Lambda dx \int dr(\alpha) :e^{i\alpha\phi_\varepsilon(x)}:.$$

As it follows from the section 2 we must to investigate the content of the "σ-algebra at infinity". To prove triviality of this σ-algebra it is sufficient to prove that

$$\lim_{\Lambda\nearrow\mathbb{R}^d} E_{\mu_\varepsilon}\{e^{i\phi(f)} \mid \Lambda^C\} = E_{\mu_\varepsilon}(e^{i\phi(f)})$$

for $f \in \mathscr{D}(\mathbb{R}^d)$. By LAMP we have

$$\begin{aligned} &E_{\mu_\varepsilon}\{e^{i\phi(f)} \mid \Lambda^C\}(\eta) \\ &= E_{\mu_\varepsilon}\{e^{i\phi(f)} \mid (\partial\Lambda)_\varepsilon\}(\eta) \\ &= E_{\mu_\Lambda}\{e^{i\phi(f)} \mid (\partial\Lambda)_\varepsilon\}(\eta) \\ &= \exp\left(i\psi_\eta^{(\partial\Lambda)_\varepsilon}(f)\right) \cdot \frac{E_0^{(\partial\Lambda)_\varepsilon}\left(e^{i\phi(\partial\Lambda)_\varepsilon(f)} e^{\lambda V_\Lambda\left(\psi_{\eta,\varepsilon}^{(\partial\Lambda)_\varepsilon} + \phi_\varepsilon^{(\partial\Lambda)_\varepsilon}\right)}\right)}{E_0^{(\partial\Lambda)}\left(e^{\lambda V_\Lambda\left(\psi_{\eta,\varepsilon}^{(\partial\Lambda)_\varepsilon} + \phi_\varepsilon^{(\partial\Lambda)_\varepsilon}\right)}\right)}. \end{aligned} \tag{5.1}$$

It follows from the Lemma 4 that

$$\lim_{\Lambda\nearrow\mathbb{R}^d} e^{i\psi_\eta^{(\partial\Lambda)_\varepsilon}(f)} = 1$$

for μ_ε-almost $\eta \in \mathscr{S}'(\mathbb{R}^d)$ and f with compact support. Therefore we need only to prove that the second factor in the last line of (5.1) tends to $E_{\mu_\varepsilon}(e^{i\phi(f)}) = J_\varepsilon(f)$ as $\Lambda \nearrow \mathbb{R}^d$. Let us consider

$$J_{\Lambda,\varepsilon}^{(\partial\Lambda)_\varepsilon,\eta}(f) = \frac{E_0^{(\partial\Lambda)_\varepsilon}\left(e^{i\phi_{(f)}^{(\partial\Lambda)_\varepsilon}} e^{-\lambda V_\Lambda\left(\psi_{\eta,\varepsilon}^{(\partial\Lambda)_\varepsilon} + \phi_\varepsilon^{(\partial\Lambda)_\varepsilon}\right)}\right)}{E_0^{(\partial\Lambda)_\varepsilon}\left(e^{-\lambda V_\Lambda\left(\psi_{\eta,\varepsilon}^{(\partial\Lambda)_\varepsilon} + \phi_\varepsilon^{(\partial\Lambda)_\varepsilon}\right)}\right)} \tag{5.2}$$

in more details. By the shift transformation

$$\Phi^{(\partial\Lambda)_\varepsilon} \to \Phi^{(\partial\Lambda)_\varepsilon} - if * S^{(\partial\Lambda)_\varepsilon} \tag{5.3}$$

we have:

$$\begin{aligned} J_{\Lambda,\varepsilon}^{(\partial\Lambda)_\varepsilon,\eta}(f) &= e^{-(1/2)\|f\|^2_{-\Lambda}(\partial\Lambda)_\varepsilon} \\ &\times \left\{1 + \sum_{n=1}^{\infty} \frac{1}{n!} \int_{\Lambda^{\otimes n}} d(x)_n \int dr(\alpha)_n \prod_{i=1}^{n} \left[e^{-\alpha_i f_\varepsilon^{(\partial\Lambda)_\varepsilon}(x_i)} - 1\right] \cdot \varrho_{\Lambda,\varepsilon}^{(\partial\Lambda)_\varepsilon,\eta}(x,\alpha)_n\right\} \end{aligned} \tag{5.4}$$

where

$$\|f\|^2_{-1,(\partial\Lambda)_\varepsilon} = \int f(x)\,(S^{(\partial\Lambda)_\varepsilon} * f)\,(x)\,dx \tag{5.4a}$$

$$f_\varepsilon^{(\partial\Lambda)_\varepsilon}(x) = \big(h_\varepsilon * (S^{(\partial\Lambda)_\varepsilon} * f)\big)\,(x) \tag{5.4b}$$

$$\varrho_{\Lambda,\varepsilon}^{(\partial\Lambda)_\varepsilon,\eta}(x,\alpha)_n = (\lambda)^n \prod_{i=1}^{n} :e^{i\alpha_i\psi_{\eta,\varepsilon}^{(\partial\Lambda)_\varepsilon}}:(x_i)\cdot\int_{\mathcal{S}'(\mathbb{R}^d)} d\mu_{\Lambda,\varepsilon}^{(\partial\Lambda)_\varepsilon}(\phi\mid\eta)\prod_{j=1}^{n} :e^{i\alpha_j\phi_{(x_j)}^{(\partial\Lambda)_\varepsilon}}: \tag{5.4c}$$

and

$$d\mu_{\Lambda,\varepsilon}^{(\partial\Lambda)_\varepsilon}(\phi\mid\eta) := \frac{1}{Z_{\Lambda,\varepsilon}^{\eta}}\exp\left\{\lambda\int_\Lambda dx\int dr(\alpha)\,:e^{i\alpha\phi_\varepsilon^{(\partial\Lambda)_\varepsilon(x)}}::e^{i\alpha\psi_{\eta,\varepsilon}^{(\partial\Lambda)_\varepsilon}}:\right\}\cdot d\mu_0^{(\partial\Lambda)_\varepsilon}\in(\Phi),$$

$$Z_{\Lambda,\varepsilon}^{\eta} := \int_{\mathcal{S}'(\mathbb{R}^d)} d\mu_0^{(\partial\Lambda)_\varepsilon}(\phi)\exp\lambda\int_\Lambda dx\int dr(\alpha)\,:e^{i\alpha\phi_\varepsilon^{(\partial\Lambda)_\varepsilon(x)}}::e^{i\alpha\psi_{\eta,\varepsilon}^{(\partial\Lambda)_\varepsilon(x)}}:. \tag{5.5}$$

The essential volume dependence of $J_{\Lambda,\varepsilon}^{(\partial\Lambda)_\varepsilon,\eta}$ comes from that of the conditioned correlation functions $\varrho_{\Lambda,\varepsilon}^{(\partial\Lambda)_\varepsilon} \equiv \{\varrho_{\Lambda,\varepsilon}^{(\partial\Lambda)_\varepsilon,\eta}\}_{n=1,\dots}$. In order to controll thermodynamic limit for them we use the formalism of the Kirkwood-Salsburg equations. By the shift transformation we have following equalities

$$\begin{aligned}\varrho_{\Lambda,\varepsilon,\eta}^{(\partial\Lambda)_\varepsilon}(x,\alpha)_n = {} & (\lambda)^n\prod_{i=1}^{n} :e^{i\alpha_i\psi^{(\partial\Lambda)_\varepsilon(x_i)}}:\cdot\exp\left(-\alpha_1\sum_{i=2}^{n}\alpha_i S_\varepsilon^{(\partial\Lambda)_\varepsilon}(x_i-x_j)\right)\\ & \times\int_{\mathcal{S}'(\mathbb{R}^d)} d\mu_{\Lambda,\varepsilon}^{(\partial\Lambda)_\varepsilon}(\Phi\mid\eta)\prod_{i=2}^{n} :e^{i\alpha_i\phi_\varepsilon^{(\partial\Lambda)_\varepsilon(x_i)}}:\\ & \times\exp\left\{\lambda\int_\Lambda dx\int dr(\alpha)\left[e^{-\alpha_1\alpha S_\varepsilon^{(\partial\Lambda)_\varepsilon(x-x_i)}}-1\right]\right.\\ & \left.\times\ :\exp i\alpha\phi_\varepsilon^{(\partial\Lambda)_\varepsilon}(x)::\exp i\alpha\psi_{\eta,\varepsilon}^{(\partial\Lambda)_\varepsilon}(x):\right\}.\end{aligned} \tag{5.6}$$

By an expansion in powers of λ we easily recognize in this equalities the Kirkwood-Salsburg equations.
Let us introduce Banach space $\mathcal{B}_\xi$ (see [*19*]) which consists of all sequences of functions $\mathcal{B}_\xi \ni \boldsymbol{\phi} = \{\phi^{(n)}(x,\alpha)_n\}$ where $\phi^n(x,\alpha)_n$ are $d(x)_n\cdot d\mu(\alpha)_n$ measurable functions and such that the norm

$$\|\boldsymbol{\phi}\|_\xi := \sup_n \frac{1}{\xi^n}\operatorname*{ess\,sup}_{(x,\alpha)_n}|\phi^{(n)}(x,\alpha)_n| \tag{5.7}$$

is finite for some positive real number ξ to be fixed below. In the space $\mathcal{B}_\xi$ we define following operator.

$$(\boldsymbol{K}_\varepsilon^{(\partial\Lambda)_\varepsilon}\boldsymbol{\phi})^{(1)}(x,\alpha)_1 = \sum_{m=1}^{\infty}\frac{1}{m!}\int d(y)_m\int dr(\beta)_m\prod_{j=1}^{m}\left[e^{-\alpha\beta_j S_\varepsilon^{(\partial\Lambda)_\varepsilon(x_1-y_j)}}-1\right]\cdot\phi^{(m)}(y,\beta)_m \tag{5.8}$$

and for $n > 1$

$$\begin{aligned}(\boldsymbol{K}_\varepsilon^{(\partial\Lambda)_\varepsilon}\boldsymbol{\phi})^{(n)}(x,\alpha)_n = {} & \exp\left(-\sum_{j=2}^{n}\alpha_i\alpha_j S_\varepsilon^{(\partial\Lambda)_\varepsilon}(x_1-x_j)\right)\\ & \times\left\{\phi^{(n-1)}(x,\alpha)'_{n-1}+\sum_{m=1}^{\infty}\frac{1}{m!}\int d(y)_m\int dr(\beta)_m\right.\\ & \left.\times\prod_{j=1}^{m}\left[e^{-\alpha_1\beta_j S^{(\partial\Lambda)_\varepsilon(x_1-y_j)}}-1\right]\Phi^{n-1+m}\big((x,\alpha)'_{n-1},(y,\beta)_m\big)\right\}.\end{aligned}$$

We define also the following operator

$$(\hat{\chi}_\Lambda \boldsymbol{\phi})^{(n)}(x, \alpha)_n = \prod_{i=1}^{n} \chi_{\Lambda_i}(x_i)\, \phi^{(n)}(x, \alpha)_n$$
$$:= \chi_\Lambda(x)_n\, \phi^{(n)}(x, \alpha)_n \tag{5.9}$$

where $\chi_\Lambda(x)_n$ is an indicator function of the set $\Lambda^{\otimes n}$. Let $\boldsymbol{a}$ be a vector from $\mathscr{B}_\xi$ given by $\alpha^{(n)} = \delta_{n,1}$. With these definitions we can rewrite the Kirkwood-Salsburg equations (5.6) in the following manner

$$\varrho_{\Lambda,\varepsilon}^{(\partial\Lambda)_\varepsilon,\eta} = -\lambda\hat{\chi}_\Lambda : e^{i\alpha_1 \psi_{\eta,\varepsilon}^{(\partial\Lambda)_\varepsilon}} : (x_1)\, \boldsymbol{a} - \lambda\hat{\chi} \boldsymbol{K}_\varepsilon^{(\partial\Lambda)_\varepsilon} \varrho_{\Lambda,\varepsilon}^{(\partial\Lambda)_\varepsilon,\eta}. \tag{5.10}$$

Introducing the permutation operator $\hat{\Pi}$ we remark that because of the symmetry of $\varrho_{\Lambda,\varepsilon}^{(\partial\Lambda)_\varepsilon}(x, \alpha)_n$ equations (5.10) hold with $\hat{\Pi}\boldsymbol{K}_\varepsilon^{(\partial\Lambda)_\varepsilon}$ instead of $\boldsymbol{K}_\varepsilon^{(\partial\Lambda)_\varepsilon}$. We choose such permutations that

$$\forall_n \sum_{i_1 \neq i_2} \alpha_{i_1}\alpha_{i_2} S_\varepsilon^{(\partial\Lambda)_\varepsilon}(x_{i_1} - x_{i_2}) \geqq \mu^2 S_\varepsilon^{(\partial\Lambda)_\varepsilon}(0).$$

We start with the following observation:

Lemma 5.1

For $\xi = C_{\zeta_\varepsilon}^{-1}$ and $|\lambda| < C_{\zeta_\varepsilon}^{-1} e^{-r^2 S_\varepsilon(0) - 1}$ we have:

$$\left\| \lambda : e^{i\alpha_1 \psi_{\eta,\varepsilon}^{(\partial\Lambda)_\varepsilon}} : (x_1)\, \hat{\Pi} \boldsymbol{K}_\varepsilon^{(\partial\Lambda)_\varepsilon} \right\|_\xi < 1$$

uniformly in $\eta \in \mathscr{S}'(\mathrm{R}^d)$.

Proof:

Let $\boldsymbol{\phi} \in \mathscr{B}_\xi$. Then

$$\frac{1}{\xi^n} \left| \left(\lambda : e^{i\alpha_1 \psi_{\eta,\varepsilon}^{(\partial\Lambda)_\varepsilon}} : (x_1)\, \hat{\Pi} \boldsymbol{K}_\varepsilon^{(\partial\Lambda)_\varepsilon} \boldsymbol{\phi}\right)^{(n)} (x, \alpha)_n \right|$$

$$\leqq \frac{1}{\xi} |\lambda| \exp(\xi C_{S_\varepsilon}) \exp\left(r^2\left(S_\varepsilon(0) - S_\varepsilon^{(\partial\Lambda)_\varepsilon}(0)\right) \exp\left(r^2 S_\varepsilon^{(\partial\Lambda)_\varepsilon}(0)\right)\right) \|\phi\|_\xi$$

$$= \frac{|\lambda|}{|\xi|} e^{r^2 S_\varepsilon(0)} \cdot \exp\left(\xi C_{S_\varepsilon}(\partial\Lambda)_\varepsilon\right) \cdot \|\phi\| \qquad \text{q.e.d.}$$

Lemma 5.2

For $\varrho_{\Lambda,\varepsilon}^{(\partial\Lambda)_\varepsilon,\eta}$ being a solution of (5.10) the norm $\left\|\varrho_{\Lambda,\varepsilon}^{(\partial\Lambda)_\varepsilon,\eta}\right\|_\xi$ is bounded uniformly in Λ and η.

Proof:

From (5.10) we have

$$\left\|\varrho_{\Lambda,\varepsilon}^{(\partial\Lambda)_\varepsilon,\eta}\right\|_\xi \leqq |\lambda| \left\| : e^{i\alpha_1 \psi_{\eta,\varepsilon}^{(\partial\Lambda)_\varepsilon}} : (x_1)\, \boldsymbol{a} \right\|_\xi + |\lambda| \left\| : e^{i\alpha_1 \psi_{\eta,\varepsilon}^{(\partial\Lambda)_\varepsilon}} : (x_1)\, \Pi \boldsymbol{K}_\varepsilon^{(\partial\Lambda)_\varepsilon} \right\|_\xi \cdot \left\|\varrho_{\Lambda,\varepsilon}^{(\partial\Lambda)_\varepsilon,\eta}\right\|_\xi.$$

Hence from the (proof of) of Lemma 5.1)

$$\left\|\varrho_{\Lambda,\varepsilon}^{(\partial\Lambda)_\varepsilon,\eta}\right\|_\xi \leqq |\lambda| \frac{1}{\xi} e^{r^2 \delta S_\varepsilon^{(\partial\Lambda)_\varepsilon}(0)} + |\lambda| \cdot \frac{1}{\xi} e^{r^2 S_\varepsilon(0) + \xi C_{S_\varepsilon}} \cdot \left\|\varrho_\Lambda^{(\partial\Lambda)_\varepsilon,\eta}\right\|_\xi.$$

Therefore

$$\left\|\varrho_{\Lambda,\varepsilon}^{(\partial\Lambda)_\varepsilon,\eta}\right\|_\xi \leqq \frac{1}{1 - \dfrac{|\lambda|}{\xi} e^{r^2 S_\varepsilon(0) + \xi C_{S_\varepsilon}}} \cdot \frac{|\lambda|}{\xi} e^{r^2 \delta S_\varepsilon^{(\partial\Lambda)_\varepsilon}(0)}.$$

For $\xi = C_{S_\varepsilon}^{-1}$ we have

$$\left\|\varrho_{\Lambda,\varepsilon}^{(\partial\Lambda)_\varepsilon,\eta}\right\|_\xi \leqq \frac{1}{1 - \frac{|\lambda|}{C_\varepsilon^{-1}} e^{r^2 S_\varepsilon(0)+1}} \cdot \frac{|\lambda|}{C_\varepsilon^{-1}} e^{r^2 S_\varepsilon(0)} \qquad \text{q.e.d.}$$

where we take $S_\varepsilon(0)$ instead of $\delta S_\varepsilon^{(\partial\Lambda)_\varepsilon}(0)$.
Crucial step in the proof of purity of the Gibbs measure $d\mu_\varepsilon$ is the following technical result:

Proposition 5.3

Let $\{\Lambda_n\}$ be a sequence of bounded sets in R^d such that $\partial\Lambda_n$ are piecewise smooth and $\Lambda_n \nearrow \mathsf{R}^d$ by inclusion and monotonously. Let $\{\Omega_n\}$ be an another sequence with the same properties and additionally $\Lambda_n \supset \Omega_n$ and $\lim\limits_{n\to\infty} \operatorname{dist}(\Omega_n, \partial\Lambda_n) = \infty$.
Then for μ_ε-almost $\eta \in \mathscr{S}'(\mathsf{R}^d)$

$$\lim_{n\to\infty} \left\|\hat{\chi}_{\Omega_n}\left(\varrho_{\Lambda_n,\varepsilon}^{(\partial\Lambda)_\varepsilon,\eta} - \varrho_\varepsilon\right)\right\|_\xi = 0 .$$

We postpone proof of this proposition to the end of this section. Hawing this result, we are in position to prove the following theorem

Theorem 5.4

If

$$|\lambda| < C_{S_\varepsilon}^{-1} e^{-r^2 S_\varepsilon(0)-1}$$

and

$$f \in \mathscr{D}(\mathsf{R}^d)$$

then

$$\lim_{\Lambda\nearrow\mathsf{R}^d} J_{\Lambda,\varepsilon}^{(\partial\Lambda)_\varepsilon,\eta}(f) = J_\varepsilon(f) .$$

The proof is based on the following sequence of Lemmas:

Lemma 5.5

There exist a constants a, $m' < m_0$ such that:

$$\left|\exp\left(-\frac{1}{2}\|f\|^2_{-1\,(\partial\Lambda)_\varepsilon}\right) - \exp\left(-\frac{1}{2}\|f\|^2_{-1}\right)\right| \leqq \|f\|_{L^2} \cdot a \cdot \exp\left(-\frac{1}{2}\|f\|^2_{-1}\right)$$
$$\times \frac{1}{2}\exp\left(-m' \operatorname{dist}(\partial\Lambda, \operatorname{supp} f)\right) \cdot |\Lambda|\, e$$

Proof:

We have from the properties of the covariances $S(x, y)$ and $S^{(\partial\Lambda)_\varepsilon}(x, y)$

$$\left|\exp\left(-\frac{1}{2}\|f\|^2_{-1,(\partial\Lambda)_\varepsilon}\right) - \exp\left(-\frac{1}{2}\|f\|^2_{-1}\right)\right|$$
$$\leqq \int_0^1 ds \exp\left(-\frac{s}{2}\|f\|^2_{-1,(\partial\Lambda)_\varepsilon}\right)\exp\left(-\frac{1-s}{2}\|f\|^2_{-1}\right) \cdot \left|\|f\|^2_{-1,(\partial\Lambda)_\varepsilon} - \|f\|^2_{-1}\right|$$
$$\leqq a' \exp\left(-\frac{1}{2}\|f\|^2_{-1}\right)\int dx\, dy f(x)\left[S(x-y) - S^{(\partial\Lambda)_\varepsilon}(x, y)\right] f(y)$$
$$\leqq a' \exp\left(-\frac{1}{2}\|f\|^2_{-1}\right) \cdot \|f\|_{L_2} \cdot \|S(x, x) - S^{(\partial\Lambda)_\varepsilon}(x, x)\|_{L_2(\operatorname{supp} f)}$$
$$\leqq \|f\|_{L_2} \exp\left(-\frac{1}{2}\|f\|^2_{-1}\right) \cdot a'|\Lambda| \exp\left(-\frac{1}{2} m' \operatorname{dist}(\partial\Lambda, \operatorname{supp} f)\right). \qquad \text{q.e d.}$$

Lemma 5.6

There exist a constant b such that:

$$\left|J_{\Lambda,\varepsilon}^{(\partial\Lambda)_{\varepsilon},\eta}(f)\right| < b \exp\left\{\xi \int dx \int d\,|r|\,(\alpha)\,|e^{-\alpha f_\varepsilon(x)} - 1|\right\}$$

Proof:

$$\left|J_{\Lambda,\varepsilon}^{(\partial\Lambda)_{\varepsilon},\eta}(f)\right| \leqq 1 + \left\|\varrho_{\Lambda,\varepsilon}^{(\partial\Lambda)_{\varepsilon},\eta}\right\|_\xi \cdot \sum_{n=1}^{\infty} \frac{\xi^n}{n!} \int dx \int d\,|r|\,(\alpha) \left(\left|\left(e^{-\alpha f_\varepsilon^{(\partial\Lambda)_\varepsilon}(x)} - 1\right)\right|\right)^n$$

$$\leqq b \exp\left\{\xi \int dx \int d\,|r|\,|(\alpha)\,|e^{-\alpha f_\varepsilon(x)} - 1|\right\}.$$

Lemma 5.7

There exists a constant $C(f, \xi)$ $a, b, m' < m_0$ such that:

$$\left|\sum_{n=1}^{\infty} \frac{1}{n!} \int d(x)_n \int dr(\alpha)_n \left[\prod_{i=1}^{n} \left(e^{-\alpha_i f_\varepsilon^{(\partial\Lambda)_\varepsilon}(x_i)} - 1\right) - \prod_{j=1}^{n} \left(e^{-\alpha_i f_\varepsilon(x_i)} - 1\right)\right] \cdot \varrho_{\Lambda,\varepsilon}^{(\partial\Lambda)_\varepsilon,\eta}(x, \alpha)_n\right|$$

$$\leqq C(f, \xi)\,(a\,|\Omega|^2\, e^{-m' \operatorname{dist}(\partial\Lambda,\Omega_\varepsilon)} + b e^{-m' \operatorname{dist}(\partial\Omega, \operatorname{supp} f)}).$$

Proof:

Using the following equality

$$\prod_{i=1}^{n} a_i - \prod_{i=1}^{n} b_i = \sum_{k=1}^{n} \prod_{i=1}^{k-1} a_i (a_k - b_k) \prod_{i=k+1}^{n} b_i$$

we have:

$$\left|\sum_{n=1}^{\infty} \frac{1}{n!} \int_{\Lambda^{\oplus n}} d(x)_n \int dr(\alpha) \left[\prod_{i=1}^{n} \left(e^{-\alpha_i f_\varepsilon^{(\partial\Lambda)_\varepsilon}(x_i)} - 1\right) - \prod_{j=1}^{n} \left(e^{-\alpha_i f_\varepsilon(x_i)} - 1\right)\right] \cdot \varrho_{\Lambda,\varepsilon}^{(\partial\Lambda)_\varepsilon,\eta}(x, \alpha)_n\right|$$

$$\leqq \left\|\varrho_{\Lambda,\varepsilon}^{(\partial\Lambda)_\varepsilon,\eta}\right\|_\xi \cdot \xi \left|\int_\Lambda dx \int dr(\alpha) \left(e^{-\alpha f_\varepsilon^{(\partial\Lambda)_\varepsilon}(x)} - e^{-\alpha f_\varepsilon(x)}\right)\right|$$

$$\times \left|\sum_{n=1}^{\infty} \frac{\xi^{n-1}}{n!} \sum_{k=1}^{n} \left[\int_\Lambda dx \int dr(\alpha) \left(e^{-\alpha f_\varepsilon^{(\partial\Lambda)_\varepsilon}(x)} - 1\right)\right]^{k-1} \cdot \left[\int_\Lambda dx \int dr(\alpha) \left(e^{-\alpha f_\varepsilon(x)} - 1\right)\right]^{n-k}\right.$$

$$\leqq C'(\xi, f) \cdot \xi \exp\left\{\xi \int dx \int dr(\alpha) \left(e^{|\alpha|\,|f_\varepsilon|(x)} - 1\right)\right\} \cdot \left|\int_\Lambda dx \int dr(\alpha)\,|\alpha|\,|f_\varepsilon^{(\partial\Lambda)_\varepsilon}(x) - f_\varepsilon(x)\right|$$

$$\leqq C''(\xi, f) \left\{\int_\Omega dx \int_{(\Omega)_\varepsilon} dy h_\varepsilon(x - y) \int dz \delta S^{(\partial\Lambda)_\varepsilon}(y - z)\,|f(z)|\right.$$

$$+ \int_{\Lambda/\Omega} dx \int_{(\Lambda/\Omega)_\varepsilon} dy h_\varepsilon(x - y) \int dz \delta S^{(\partial\Lambda)_\varepsilon}(y - z)\,|f(z)|$$

$$\leqq C''(\xi, f) \left\{|\Omega|\,|(\Omega)_\varepsilon| \sup_{x \in \Omega_\varepsilon} |\delta S^{(\partial\Lambda)_\varepsilon}(x)| \int dz\,|f(z)| + \int_{(\Lambda/\Omega)_\varepsilon} dy \sup_{z \in \operatorname{supp} f} |\delta S^{(\partial\Lambda)_\varepsilon}(y - z)| \left(\int dz\,|f(z)|\right)\right\}$$

$$\leqq C(f, \xi)\,(a\,|\Omega|^2\, e^{-m' \operatorname{dist}(\partial\Lambda,\Omega_\varepsilon)} + b e^{-m' \operatorname{dist}(\partial\Omega, \operatorname{supp} f)}).$$

Lemma 5.8

There exists a constant $D(f, \xi)$ and $m' < m_0$ such that:

$$\left|\frac{1}{n!} \int d(x)_n \int dr(\alpha)_n \prod_{j=1}^{n} \left[e^{-\alpha_j f_\varepsilon(x_i)} - 1\right] (1 - \chi_\Omega(x)_n) \left(\chi_\Lambda(x)_n\right) \cdot \left(\varrho_{\Lambda,\varepsilon}^{(\partial\Lambda)_\varepsilon,\eta} - \varrho_\varepsilon\right)(x, \alpha)_n\right|$$

$$\leqq D(f, \xi)\, e^{-m' \operatorname{dist}(\operatorname{supp} f, \partial\Omega_\varepsilon)}.$$

Proof:

$$\left|\sum_{n=1}^{\infty} \frac{1}{n!} \int d(x)_n \int dr(\alpha)_n \prod_{j=1}^{n} [e^{-\alpha_j f_\varepsilon(x_j)} - 1] (1 - \chi_\Omega)(x)_n \cdot \chi_\Lambda(x_n) \cdot \left(\varrho_{\Lambda,\varepsilon}^{(\partial\Lambda)_{\varepsilon,\eta}} - \varrho_\varepsilon\right)(x, \alpha)_n\right|$$
$$\leqq \left(\left\|\varrho_{\Lambda,\varepsilon}^{(\partial\Lambda)_{\varepsilon,\eta}}\right\|_\xi + \|\varrho_\varepsilon\|_\xi\right) \cdot \exp\left(\xi \int dx \int d|r|(\alpha)\,(e^{|\alpha||f|_\varepsilon(x)} - 1)\right) \cdot \int_{\Omega^C} dx \int d|r|(\alpha)\,(e^{|\alpha||f_\varepsilon|(x)} - 1)$$
$$\leqq D'(f, \xi) \int_{\Omega^C} dx \int d|r|(\alpha)\,(e^{|\alpha||f_\varepsilon|(x)} - 1)$$
$$\leqq D''(f, \xi) \int_{\Omega^C} dx\, |f_\varepsilon|(x)$$
$$= D''(f, \xi) \int_{\Omega^C} dx \int_{(\Omega)_\varepsilon{}^C} dy h_\varepsilon(x - y) \int dz S(y - z)\, |f(z)|$$
$$\leqq D'''(f, \xi) \int_{\Omega^C} dy \sup_{z \in \mathrm{supp} f} S(y, z)$$
$$\leqq D(f, \xi)\, e^{-m' \mathrm{dist}(f, (\partial\Omega)_\varepsilon)}.$$

Proof of Theorem 5.4

We have:

$$\left|J_{\Lambda,\varepsilon}^{(\partial\Lambda)_{\varepsilon,\eta}}(f) - J_\varepsilon(f)\right|$$
$$= \left(\exp\left(-\frac{1}{2}\|f\|_{-1,(\partial\Lambda)_\varepsilon}^2\right) - \exp\left(-\frac{1}{2}\|f\|_{-1}^2\right) \cdot \exp\left(\frac{1}{2}\|f\|_{-1,(\partial\Lambda)_\varepsilon}^2\right)\right) \cdot J_{\Lambda,\varepsilon}^{(\partial\Lambda)_{\varepsilon,\eta}}(f)$$
$$+ \exp\left(-\frac{1}{2}\|f\|_{-1}^2\right) \cdot \sum_{n=1}^{\infty} \frac{1}{n!} \int_{\Lambda^{\otimes n}} d(x)_n \int dr(\alpha)_n \left[\prod_{i=1}^{n} \left(e^{-\alpha_i f_\varepsilon^{(\partial\Lambda)_\varepsilon}(x_i)} - 1\right) - \prod_{j=1}^{n} (e^{-\alpha_i f_\varepsilon(x_i)} - 1)\right]$$
$$\times\, \varrho_{\Lambda,\varepsilon}^{(\partial\Lambda)_{\varepsilon,\eta}}(x, \alpha)_n$$
$$+ \exp\left(-\frac{1}{2}\|f\|_{-1}^2\right) \cdot \sum_{n=1}^{\infty} \frac{1}{n!} \int d(x)_n \int dr(\alpha)_n \prod_{j=1}^{n} (e^{-\alpha_i f_\varepsilon(x_i)} - 1)\, \chi_\Omega(x)_n$$
$$\times \left(\varrho_{\Lambda,\varepsilon}^{(\partial\Lambda)_{\varepsilon,\eta}}(x, \alpha)_n - \varrho_\varepsilon(x, \alpha)_n\right)$$
$$+ \exp\left(-\frac{1}{2}\|f\|_{-1}^2\right) \cdot \sum_{n=1}^{\infty} \frac{1}{n!} \int d(x)_n \int dr(\alpha)_n \prod_{j=1}^{n} (e^{-\alpha_i f_\varepsilon(x_i)} - 1)\,(1 - \chi_\Omega)(x)_n$$
$$\times\, \chi_\Lambda(x)_n \left(\varrho_{\Lambda,\varepsilon}^{(\partial\Lambda)_{\varepsilon,\eta}}(x, \alpha)_n - \varrho(x, \alpha)_n\right).$$

Each of the terms is estimated in the listed sequence of Lemmas. Hence

$$\lim_{\Lambda \nearrow \mathbf{R}^d} J_{\Lambda,\varepsilon}^{(\partial\Lambda)_{\varepsilon,\eta}}(f) = J_\varepsilon(f). \qquad \text{q.e.d.}$$

From the Theorem 5.4 we have proof of the theorem 1.1 from the introduction.
We proceed now to the proof of Proposition 5.3.
Let $\boldsymbol{K}_\varepsilon$ be $K - S$ operator for ϱ_ε. It differs from $\boldsymbol{K}_\varepsilon{}^{(\partial\Lambda)_\varepsilon}$ only in that in place of $S_\varepsilon{}^{(\partial\Lambda)_\varepsilon}$ we have always S_ε. We have the following lemma.

Lemma 5.9

Let $\{\Lambda_n\}$, $\{\Omega_n\}$ be a sequences as in Proposition 5.3. Let $\{\Omega_n'\}$ be a sequence with the properties as $\{\Lambda_n\}$ and additionally

$$\forall_n : \Omega_n \subset \Omega'_{n'} \subset \Lambda_n$$

and

$$\lim_{n\to\infty} \operatorname{dist}(\Omega_n, \partial\Omega_n') = \infty, \quad \lim_{n\to\infty} \operatorname{dist}(\Omega_n', \partial\Lambda_n) = \infty.$$

Then:

$$\lim_{n\to\infty} \|\hat{\chi}_{\Omega_n}(\hat{\Pi}\boldsymbol{K}_\varepsilon^{(\partial\Lambda)_\varepsilon} - \hat{\Pi}\boldsymbol{K}_\varepsilon)\,\hat{\chi}_{\Omega_n'}\|_\xi = 0.$$

Proof:

Let $\boldsymbol{\phi} \in \mathscr{B}_\xi$. We have from the definitions of the K-S operators

$$\begin{aligned}
&|(\hat{\chi}_{\Omega_n}(\hat{\Pi}\boldsymbol{K}_\varepsilon^{(\partial\Lambda_n)_\varepsilon} - \Pi\boldsymbol{K}_\varepsilon)\,\hat{\chi}_{\Omega_n}\cdot\boldsymbol{\phi})^{(k)}(x,\alpha)_k| \\
&\leqq \|\boldsymbol{\phi}\|_\xi\cdot\xi^{k-1}\exp(\xi C_{S_\varepsilon}^{-1}) \\
&\quad\times\left|\chi_{\Omega_n}(x)_k\left(\exp\left(-\sum_{j=2}^k s_1s_jS_\varepsilon^{(\partial\Lambda)_\varepsilon}(x_1,x_j)\right) - \exp\left(-\sum_{j=2}^k s_1s_jS_\varepsilon(x_i - x_j)\right)\right)\right| \\
&\quad+\|\boldsymbol{\phi}\|_\xi\cdot\xi^{k-1}\exp\left(-\sum_{j=2}^k s_1s_jS_\varepsilon(x_1,x_j)\right)\cdot\left|\chi_{\Omega_k}(x)\cdot\sum_{m=1}^\infty\frac{\xi^m}{m!}\int_{\Omega_n'} d(y)_m\int dr(\beta)_m\right. \\
&\quad\left.\times\sum_{i=1}^m\prod_{j=1}^{i-1}\left(e^{-\alpha_1\beta_jS_\varepsilon^{(\partial\Lambda)_\varepsilon}(x_1-y_j)}\right)\cdot\left(e^{-\alpha_1\beta_iS_\varepsilon^{(\partial\Lambda)_\varepsilon}(x_1,y_i)} - e^{-\alpha_1\beta_jS_\varepsilon(x_1,y_i)}\right)\cdot\prod_{j=i+1}^m e^{-\alpha_1\beta_jS_\varepsilon(x-y_j)}\right| \\
&\leqq \|\boldsymbol{\phi}\|_\xi\cdot\xi^{k-1}\left\{\left|\chi_{\Omega_k}(x)_k\right|\sum_{i=2}^k\alpha_1\alpha_i\delta S_\varepsilon^{(\partial\Lambda)_\varepsilon}(x_1,x_i)\right| \\
&\quad\left.\times 2e^{(k-1)S_\varepsilon(0)}\exp(\xi C_{S_\varepsilon}^{-1}) + ae^{(k-1)r^2S_\varepsilon(0)}\xi r\exp(\xi C_{S_\varepsilon}^{-1})\cdot e^{-m'\operatorname{dist}(\Omega_n',\partial\Lambda_n)}\right\} \\
&\leqq \|\boldsymbol{\phi}\|_\xi\cdot\xi^{k-1}\left\{2(k-1)\,r^2a_1e^{-m'\operatorname{dist}(\Omega_n,\partial\Lambda_n)} + \xi re^{-m'\operatorname{dist}(\Omega_n',\partial\Lambda_n)}\right. \\
&\quad\left.\times\frac{1}{\xi}\exp\left((k-1)\,r^2S_\varepsilon(0) + \xi C_{S_\varepsilon}^{-1}\right)\right\}.
\end{aligned}$$

where a, a_1, $m < m_0$ are some constants.
The last estimate proves that every component of the vector $\left(\hat{\chi}_{\Omega_n}(\hat{\Pi}\boldsymbol{K}_\varepsilon^{(\partial\Lambda_n)_\varepsilon} - \hat{\Pi}\boldsymbol{K}_\varepsilon)\times\hat{\chi}_{\Omega n'})\,\boldsymbol{\phi}\right)$ tends uniformly to zero as $n\to\infty$.
Moreover because

$$\chi_{\Omega_n}(x)\,\delta S_\varepsilon^{(\partial\Lambda_n)_\varepsilon}(x,x')\,\chi_{\Omega_n}(x')\to 0$$

as $\operatorname{dist}(\partial\Lambda_n,\Omega_n)\to\infty$ it follows that if

$$\sum_{j=2}^k\alpha_1\alpha_jS_\varepsilon(x_1,x_j) \geqq -r^2S_\varepsilon(0)$$

then also

$$\sum_{j=2}^k\alpha_1\alpha_jS_\varepsilon^{(\partial\Lambda)_\varepsilon}(x_1,x_j) \geqq -r^2S_\varepsilon^{(\partial\Lambda)_\varepsilon}(0)$$

for sufficiently large $\operatorname{dist}(\Omega_n,\partial\Lambda_n)$.
This fact proves that the limit

$$\lim_{n\to\infty}\|\hat{\chi}_{\Omega_n}(\hat{\Pi}\boldsymbol{K}_\varepsilon^{(\partial\Lambda_n)_\varepsilon} - \hat{\Pi}\hat{\boldsymbol{K}}_\varepsilon)\,\hat{\chi}_{\Omega_n'}\boldsymbol{\phi}\|_\xi$$

exists. Hence

$$\lim_{n\to\infty}\|\chi_{\Omega_n}(\hat{\Pi}\hat{\boldsymbol{K}}_\varepsilon^{(\partial\Lambda_n)_\varepsilon} - \hat{\Pi}\boldsymbol{K}_\varepsilon)\,\chi_{\Omega_n'}\|_\xi = 0 \qquad \text{q.e.d.}$$

Lemma 5.10

Let $\{\Lambda_n\}$, $\{\Omega_n'\}$, $\{\Omega_n\}$ be as in the Lemma 5.9. Then

$$\lim_{n\to\infty} \|\hat{\chi}_{\Omega_n} \boldsymbol{K}_\varepsilon^{(\partial\Lambda_n)_\varepsilon}(1 - \hat{\chi}_{\Omega_n'})\|_\xi = 0$$

and

$$\lim_{n\to\infty} \|\chi_{\Omega_n} \boldsymbol{K}_\varepsilon(1 - \hat{\chi}_{\Omega_n'})\|_\xi = 0.$$

Proof:

For $\boldsymbol{\phi} \in \mathscr{B}_\xi$ we have

$$\begin{aligned}
&\left|\left(\hat{\chi}_{\Omega_n} \boldsymbol{K}_\varepsilon^{(\partial\Lambda_n)_\varepsilon}(1 - \hat{\chi}_{\Omega_n'})\, \boldsymbol{\phi}\right)^{(1)}(x, \alpha)\right| \\
&= \left|\left(\hat{\chi}_{\Omega_n}(x)\, \boldsymbol{K}_\varepsilon^{(\partial\Lambda_n)_\varepsilon}\left(1 - \hat{\chi}_{\Omega_n}(x)\right) \phi\right)^{(1)}(x, \alpha)\right| \\
&= \left|\chi_{\Omega_n}(x) \cdot \sum_{m=1}^{\infty} \frac{1}{m!} \int_{(\Omega_n')^C} d(y)_m \int dr(\beta)_m \prod_{j=1}^{m} \left(e^{-\alpha_1\beta S_\varepsilon^{(\partial\Lambda)_\varepsilon}(x_1-y)} - 1\right) \phi^{(m)}(y, \beta)_m\right| \\
&\leqq a\, \|\boldsymbol{\phi}\|_\xi \exp\left(\xi C_{S_\varepsilon}\right) \exp\left(-m' \operatorname{dist}\left(\Omega_n, \partial\Omega_n'\right)\right).
\end{aligned}$$

If $n > 1$ then

$$\left(\hat{\chi}_{\Omega_n} \mathbf{K}_\varepsilon^{(\partial\Lambda)_\varepsilon}(1 - \hat{\chi}_{\Omega_n'})\, \boldsymbol{\phi}\right)^{(n)}(x, \alpha)_n = 0.$$

The same is true for $\boldsymbol{K}_\varepsilon$. q.e.d.

Both ϱ_ε and $\varrho_{\Lambda,\varepsilon}^{(\partial\Lambda)_{\varepsilon,\eta}}$ are solutions of the K-S type equations. Therefore they a given (for small λ) by strongly convergent (in $\mathscr{B}_\xi$) expansions:

$$\varrho_{\Lambda,\varepsilon}^{(\partial\Lambda)_{\varepsilon,\eta}} = -\lambda \sum_{m=0}^{\infty} (-\lambda\hat{\chi}_\Lambda \hat{\Pi} \boldsymbol{K}_\varepsilon^{(\partial\Lambda)_\varepsilon})^m \left(\chi_\Lambda(x)_1 : e^{i\alpha_1\psi_{\eta,\varepsilon}^{(\partial\Lambda)_\varepsilon}} : (x_1)\, \hat{\alpha}\right)$$

and

$$\varrho_\varepsilon = -\lambda \sum_{m=0}^{\infty} (-\lambda\hat{\Pi}\boldsymbol{K}_\varepsilon)^m\, \alpha.$$

We have:

Lemma 5.11

For any $m = 1, 2, \ldots$

$\{\Lambda_n\}$, $\{\Omega_n\}$ as in Lemma 5.9

$$\lim_{n\to\infty} \left\|\hat{\chi}_{\Omega_n} \left[(\hat{\chi}_{\Lambda_n} \hat{\Pi} \boldsymbol{K}_\varepsilon^{(\partial\Lambda)_\varepsilon})^m\, \hat{\chi}_{\Lambda_n} : e^{i\alpha_1\psi_{\eta,\varepsilon}^{(\partial\Lambda_n)_\varepsilon}} : (x_1)\, \alpha - (\Pi\hat{\boldsymbol{K}}_\varepsilon)^m\, \alpha\right]\right\|_\xi = 0.$$

Proof:

Let us consider m-sequences $\{\Omega_n^{(i)}\}$ $i = 1, 2, \ldots, m$ such that $\forall_{k,i}\ \Omega_k^{(i)} \subset \Omega_k^{(i+1)}$,

$$\operatorname{dist}(\Omega_n, \partial\Omega_n^{(1)}), \ldots, \operatorname{dist}(\Omega_n^{(i)}, \partial\Omega_n^{(i+1)}), \ldots, \operatorname{dist}(\Omega_n^{(m)}, \partial\Lambda_n)$$

tends to infinity as $n \to \infty$ and further properties as above.

Using this we can write:

$$\begin{aligned}
&\hat{\chi}_{\Omega_n}(\hat{\chi}_{\Lambda_n} \hat{\Pi} \boldsymbol{K}_\varepsilon^{(\partial\Lambda)_\varepsilon})^m\, \chi_{\Lambda_n} : e^{i\alpha_1\psi_{\eta,\varepsilon}^{(\partial\Lambda_n)_\varepsilon}} : (x_1)\, \alpha - \hat{\chi}_{\Omega_n}(\hat{\Pi}\boldsymbol{K}_\varepsilon)^m\, \alpha \\
&= \hat{\chi}_{\Omega_n}(\hat{\chi}_{\Lambda_n} \hat{\Pi} \boldsymbol{K}_\varepsilon^{(\partial\Lambda_n)_\varepsilon}) \prod_{i=1}^{m} \left[\left(\hat{\chi}_{\Omega_n^{(i)}} + (1 - \chi^i)_{\Omega_n^{(i)}}\right) \hat{\Pi} \boldsymbol{K}^{(\partial\Lambda_n)}\right. \\
&\quad \times \left.(\hat{\chi}_{\Omega_n^{(i+1)}} + 1 - \hat{\chi}_{\Omega_n^{(i+1)}})\right] \hat{\chi}_{\Lambda_n} : e^{i\alpha_1\psi_{\eta,\varepsilon}^{(\partial\Lambda)_\varepsilon}} : \alpha - \hat{\chi}_{\Omega_n}(\hat{\Pi}\boldsymbol{K}_\varepsilon)^m\, \alpha.
\end{aligned}$$

Performing the multiplication of the characteristic functions we arrive to the situation where the Lemmas 5.10; 5.9, and Lemma 4.2 can be applied.

Proof of the Proposition 5.3

Having proven Lemma 5.11 we proceed further as in the proof of the Th. 4.2.3 in RUELLE [19]. This argumentation gives

$$\lim_{n\to\infty} \|\chi_{\Omega_n}\{[\hat{1} - \lambda\hat{\chi}_\Lambda \hat{\Pi}\mathbf{K}_\varepsilon^{(\partial\Lambda_n)_\varepsilon}]^{-1} - [1 - \lambda\hat{\chi}_\Lambda \Pi \mathbf{K}_\varepsilon]^{-1}\|_\xi = 0 \qquad \text{q.e.d.}$$

6. Global Almost Markov Property

Let C^0 be a smooth hypersurface which divide R^d into two disjoint sets Ω^C (open) and Ω. Let $f \in \mathcal{S}(\mathsf{R}^d)$ with compact support in Ω. Let $\{\Lambda_n\}$ be a sequence of subsets of R^d such that: $\operatorname{supp} f \subset \Lambda_n \cap \Omega$, $\partial\Lambda_n$ are smooth hypersurfaces such that $\partial\Lambda_n \cap C^0$ are $(d-2)$-dimensional. The measure $d\mu_\varepsilon$ has LAMP, therefore we have:

$$\begin{aligned} &E_{\mu_\varepsilon}\{e^{i\phi(f)} \mid (\Lambda_n{}^C + \Omega^C)_\varepsilon\} \\ &= E_{\mu_\varepsilon}\{e^{i\phi(f)} \mid (\partial\Lambda_n \cup C^0)_\varepsilon\} \\ &= E_{\mu_{\Lambda_n}}\{e^{\phi(f)} \mid (\partial\Lambda_n \cup C^0)_\varepsilon\}. \end{aligned} \tag{6.1}$$

We show the following result.

Proposition 6.1

Let C^0, $\{\Lambda_n\}$, f be as above. Then

$$\lim_{n\to\infty} E_{\mu_\varepsilon,\Lambda_n}\{e^{i\phi(f)} \mid (C^0 \cup \partial\Lambda_n)_\varepsilon\} = E_{\mu_\varepsilon}\{e^{i\phi(f)} \mid (C^0)_\varepsilon\} \tag{6.2}$$

almost sure $w.\ r$ to $d\mu_\varepsilon$

Proof:

We need to show that the measure $\mu_\varepsilon{}^{C^0}(\phi \mid \eta)$ given by its characteristic functional $E_{\mu_\varepsilon}\{e^{i\phi(f)} \mid (C^0)_\varepsilon\}$ is a pure Gibbs measure corresponding to the almost-additive functional

$$V_{\Lambda,\iota,\eta}(\phi) = \lambda \int_\Lambda dx \int dr(\alpha) :e^{i\alpha\phi^{(C^0)_\varepsilon}}: :e^{i\alpha\psi_{\eta_\varepsilon}^{(C_0)_\varepsilon}}: \tag{6.3}$$

The method of the proof is the same as in the previous section. We represent $E_{\mu_{\Lambda_n},\varepsilon}\{e^{i\phi(f)} \mid (C^0 \cup \partial\Lambda_n)_\varepsilon\}$ and $E_{\mu_\varepsilon}\{e^{i\phi(f)} \mid (C^0)_\varepsilon\}$ in terms of the conditioned correlation functions $\varrho_{\Lambda_n,\varepsilon,\eta}^{(C_0 \cup \partial\Lambda_n)_\varepsilon}$ and $\varrho_\varepsilon{}^{(C_0)_\varepsilon}$ respectively to which we apply the K-S type of analysis. Then we prove that $\varrho_{\Lambda_n,\varepsilon,\eta}^{(C^0 \cup \partial\Lambda_n)_\varepsilon}$ tends locally compactly to $\varrho_\varepsilon{}^{(C^0)_\varepsilon}$ as $\Lambda_n \nearrow \mathsf{R}^d$ for almost every η $w.\ r$ to $d\mu_\varepsilon$.

As in sections 3 and 5 we have

$$\begin{aligned} &E_{\mu_{\Lambda_n,\varepsilon}}\{e^{i\phi(f)} \mid (C^0 \cup \partial\Lambda_n)_\varepsilon\}\,(\eta) \\ &= \exp\left(-\frac{1}{2}\|f\|_{-1,}^{2}(C^0 \cup \partial\Lambda_n)\right)\varepsilon \cdot \exp\,(i\psi_\eta{}^{(C^0 \cup \partial\Lambda_n)_\varepsilon})\,(f) \\ &\quad \times \int_{\mathcal{S}'(\mathsf{R}^d)} \exp\Big\{\lambda \int_\Lambda dx \int dr(\alpha) :\exp\left(i\alpha\phi_{(x)}^{(C_0 \cup \partial\Lambda_n)_\varepsilon}\right): :\exp\left(i\alpha\psi_\eta{}^{(C_0 \cup \partial\Lambda_n)_\varepsilon}\right): \\ &\quad \times \left(\exp\left(-\alpha f_\varepsilon{}^{(C^0 \cup \partial\Lambda_n)_\varepsilon}(x_j) - 1\right)\right)\Big\}\, d\mu_{\Lambda_n}^{(C^0 \cup \partial\Lambda_n)_\varepsilon}(\phi \mid \eta). \end{aligned} \tag{6.4}$$

The conditioned correlation functions

$$\varrho_{\Lambda_n,\varepsilon,\eta}^{(C_0\cup\partial\Lambda_n)_\varepsilon}(x,\alpha)_n = \prod_{i=1}^n :e^{i\alpha_i\psi_\eta^{(C_0\cup\partial\Lambda_n)_\varepsilon(x)}}: \int_{\mathcal{S}'(\mathbf{R}^d)} \prod_{i=1}^n :e^{i\alpha\phi^{(C^0\cup\partial\Lambda_n)_\varepsilon(x_i)}}: d\mu_{\Lambda_n,\varepsilon}^{(C^0\cup\partial\Lambda_n)_\varepsilon}(\phi \mid \eta) \quad (6.5)$$

fulfill the K-S equations:

$$\varrho_{\Lambda_n,\varepsilon,\eta}^{(C^0\cup\partial\Lambda_n)_\varepsilon} = -\lambda\hat{\chi}_{\Lambda_n} :e^{i\alpha_1\psi^{(C^0\cup\partial\Lambda_n)_\varepsilon(x_1)}}: \alpha - \lambda\hat{\chi}_{\Lambda_n}\hat{\Pi}\boldsymbol{K}_\varepsilon^{(C^0\cup\partial\Lambda_n)} \cdot \varrho_{\Lambda_n,\varepsilon,\eta}^{(C^0\cup\partial\Lambda_n)_\varepsilon}. \quad (6.6)$$

We have also

$$E_{\mu_\varepsilon}\{e^{i\phi(f)} \mid (C^0)_\varepsilon\}(\eta)$$
$$= \exp\left(-\frac{1}{2}\|f\|^2_{-1,(C^0)_\varepsilon}\right)$$
$$\times \int_{\mathcal{S}'(\mathbf{R}^d)} \exp\left\{\lambda \int_\Lambda dx \int dr(\alpha)\left(e^{-\alpha f_\varepsilon C_\varepsilon^0(x)} - 1\right) \cdot :e^{i\alpha\phi(C)_\varepsilon^0(x)}: :e^{i\alpha\psi_\eta C_\varepsilon^0}:(x)\right\} d\mu_\varepsilon^{(C^0)_\varepsilon}(\phi \mid \eta). \quad (6.7)$$

The corresponding here correlation functions $\varrho_{\varepsilon,\eta}^{(C_0)_\varepsilon}$ fulfil the K-S equations

$$\varrho_{\varepsilon,\eta}^{(C_0)_\varepsilon} = -\lambda :e^{i\alpha_1\psi_\eta^{(C^0)_\varepsilon}}:(x_1)\,\alpha - \lambda\hat{\Pi}\boldsymbol{K}_\varepsilon^{(C^0)_\varepsilon}\varrho_{\varepsilon,\eta}^{(C^0)_\varepsilon}. \quad (6.8)$$

For

$$|\lambda| < C_{S_\varepsilon(0)}\, e^{-\tau^2 S_\varepsilon(0)-1}.$$

Both (6.6) and (6.8) have unique solutions in the space $\mathcal{B}_\xi$ given by the strongly convergent Liouville-Neumann expansions of the corresponding operators. The following generalization of the Proposition 5.3 can be proved identically as the original one:

Lemma 6.2

Let $\{\Lambda_n\}$ be a sequence of bounded sets in $\mathbf{R}^d$ as in proposition (6.1). Let $\{\Omega_n\}$ be another sequence fulfilling the assumptions of Proposition 5.3. Then for μ_ε-almost all $\eta \in S'(\mathbf{R}^d)$

$$\lim_{n\to\infty} \left\|\chi_{\Omega_n}\left(\varrho_{\Lambda_n,\varepsilon,\eta}^{(C^0\cup\partial\Lambda_n)_\varepsilon} - \varrho_{\varepsilon,\eta}^{(C^0)_\varepsilon}\right)\right\|_\xi = 0.$$

From this and properties of the $S^{(C_0\cup\partial\Lambda_n)_\varepsilon} - S^{(C_0)_\varepsilon}$ we obtain identically as in the section 5 the proof of the statement of the Proposition 6.1. q.e.d.

Now we are ready to prove Theorem 1.2 from the Introduction.

Proof of Theorem 1.2

From the definition of $E_{\mu_\varepsilon}\{\cdot \mid (C^0)_\varepsilon\}$ we see that it depends of η only through the field $\psi_\eta^{(C^0)_\varepsilon}$ which by definition is $\Sigma_{(C^0)_\varepsilon}$ measurable. Hence the functions

$$\eta \to \int e^{i\phi(f)}\, d\mu_{\Lambda_n,\varepsilon}^{(C^0\cup\partial\Lambda_n)_\varepsilon}(\phi \mid \eta)$$

are all $\Sigma_{(C_0)_\varepsilon}$ measurable. On the other hand we have

$$e^{i\psi_\eta^{(C^0\cup\partial\Lambda_n)_\varepsilon(f)}} \int e^{i\phi(f)}\, d\mu_{\Lambda_n,\varepsilon}^{(C^0\cup\partial\Lambda_n)_\varepsilon}(\phi \mid \eta)$$
$$= e^{i\psi_\eta^{(C^0)_\varepsilon(f)}}\, e^{i\left(\psi_\eta^{(C^0\cup\partial\Lambda_n)_\varepsilon(f)} - \psi_\eta^{(C^0)_\varepsilon(f)}\right)} \cdot E_{\mu_\varepsilon}\{e^{i\phi(f)} \mid (C^0 \cup \partial\Lambda_n)_\varepsilon\}.$$

By the conditional expectation $E_{\mu_\varepsilon}\{\cdot \mid (C^0 \cup \mathbf{R}^2 \setminus \Lambda_n)_\varepsilon\}$ is by definition a martingale in n so by the martingale convergence theorem we have that

$$\lim_{n\to\infty} E_{\mu_\varepsilon}\{\cdot \mid (C^0 \cup \mathbf{R}^2 \setminus \Lambda_n)_\varepsilon\}(\eta) \quad \text{exists for}$$

almost every η with respect to $d\mu_\varepsilon$. Hence it follows that $\lim_{n\to\infty} \int e^{i\phi(f)}\, d\mu_{\Lambda_n(\phi|\eta)}^{(C^0\cup\partial\Lambda_n)_\varepsilon}$ converges for almost all η.

This limit is equal to $E_{\mu_\varepsilon}\left\{e^{i(\phi-\varphi_\eta^{(C_0)_\varepsilon})(f)} \,\middle|\, \Sigma^\infty_{(C^0)_\varepsilon}\right\}$ where

$$\Sigma^\infty_{(C_0)_\varepsilon} = \bigcap_n \Sigma_{(C^0\cup\Lambda_n{}^c)_\varepsilon}$$

Taking into account that

$$E_{\mu_\varepsilon}\left\{e^{i(\phi-\varphi_\eta^{(C_0)_\varepsilon})(f)} \,\middle|\, \Sigma^\infty_{(C^0)_\varepsilon}\right\}(\eta) = e^{i\varphi_\eta^{(C^0)_\varepsilon(f)}} \cdot E_{\mu_\varepsilon}\{e^{i\phi(f)} \mid \Sigma^\infty_{(C_0)_\varepsilon}\}(\eta)$$

we conclude that $E_{\mu_\varepsilon}\{\exp i\phi(f) \mid \Sigma^\infty_{(C_0)_\varepsilon}\}$ is $\Sigma_{(C_0)_\varepsilon}$ measurable as a function of η. This yields that for every $F\ L^\infty(d\mu_\varepsilon)$, $E_{\mu_\varepsilon}\{F \mid \Sigma^\infty_{(C^0)_\varepsilon}\}$ is $\Sigma_{(C^0)_\varepsilon}$ measurable.
Let $\mathsf{R}^d \setminus (C^0)_\varepsilon = \Omega_+ \cup \Omega_-$ where $\Omega_\pm$ are connected components of the set $\mathsf{R}^d \setminus (C^0)_\varepsilon$ and let F_+ and F_- be bounded functions which are $\Sigma_{\Omega^+} \cap \Sigma_\Lambda$ resp. $\Sigma_{\Omega^-} \cap \Sigma_\Lambda$ measurable. By the discussion above we have

$$E_{\mu_\varepsilon}\{F_+F_- \mid (C_0)_\varepsilon\}(\eta)$$

(by Prop. 6.1) $$= \lim_{n\to\infty} E_{\mu_\varepsilon}\{F_+F_- \mid (C_0 \cup \mathsf{R}\setminus \Lambda_n)_\varepsilon\}(\eta)$$

(by LAMP) $$= \lim_{n\to\infty} E_{\mu_\varepsilon}\{F_+F_- \mid (C^0 \cup \partial\Lambda_n)_\varepsilon\}(\eta)$$

(by LAMP) $$= \lim_{n\to\infty} E_{\mu_\varepsilon}\{F_+ \mid (C^0 \cup \partial\Lambda_n)_\varepsilon\}(\eta)\, E_{\mu_\varepsilon}\{F_- \mid (C^0 \cup \partial\Lambda_n)_\varepsilon\}(\eta)$$

(by Prop. 6.1) $$= \lim_{n\to\infty} E^{(C_0)_\varepsilon}_{\mu_\varepsilon}\{F_+ \mid (C^0)_\varepsilon\}(\eta)\, E_{\mu_\varepsilon}\{F_- \mid (C^0)\}(\eta)$$

which is exactly the GAMP

q.e.d.

Remarks

It is possible to choose the regularization local in time, say the first component x_1 of $x\ \mathsf{R}^d$. Using such regularization we obtain Markov property for all hypersurfaces $x_1 = \text{const}$. This leads to the existence of selfadjoint transfer matrix in the time direction i.e. the existence of selfadjoint Hamiltonian in the corresponding to this measure quantum field theory [*29*]. Because the regularization is local in time it does'nt destroy the canonical (equal-time) commutation relations. From the Markov property for constant times hypersurfaces it follows that the physical vacuum is an cyclic vector for the time-zero quantum field i.e. the corresponding field theory is canonical in the sense of ARAKI [*30*].

References

[*1*] DOBRUSHIN, R. L., Funct. Anal. Applic. **2** (1968) 44.
[*2*] DOBRUSHIN, R. L., Theory Probab. Appl. **15** (1970) 469.
[*3*] LANFORD, O. E., RUELLE, D., Commun. Math. Phys. **13** (1969) 194.
[*4*] MINLOS, R. A., Funct. Anal. Applic. **1** (1967) 60.
[*5*] MINLOS, R. A., Funct. Anal. Applic. **1** (1967) 40.
[*6*] SINAI, YA. G., Theory of Phase Transitions, Moscow "Science", 1980, in Russian.
[*7*] RUELLE, D., Helv. Phys. Acta **36** (1963) 187.
[*8*] SIEGERT, A. J. F., Physica **26** (1960) 30.
[*9*] FRÖHLICH, J., PARK, X. M., Commun. Math. Phys. **59** (1978) 235.
[*10*] FRÖHLICH, J., Commun. Math. Phys. **47** (1976) 233.

[11] FRÖHLICH, J., SEILER, E., Helv. Phys. Acta **49** (1976) 889.
[12] FRÖHLICH, J., PARK, Y. M., IHES, preprint, 1980.
[13] ALBEVERIO, S., HOEGH-KRÖHN, R., Commun. Math. Phys. **68** (1980) 95.
[14] Constructive Q.F. Theory (eds. G. VELO, A. S. WIGHTMAN), Lecture Notes in Physics, Vol. 25, Berlin—Heidelberg—New York, Springer 1973.
[15] MOLCZAN, G. M., Dokl. Akad. Nauk SSSR **215** (1974), No. 5.
[16] MOLCZAN, G. M., Dokl. Akad. Nauk SSSR **197** (1971) No. 4.
[17] PITT, D., Arch. Ration, Mech. Anal. **43** (1971) 5.
[18] GIELERAK, R., ZEGARLINSKI, B., (in preparation).
[19] RUELLE, D., Statistical Mechanics. Rigorous Results, New York: Benjamin 1969.
[20] SIMON, B., The $P(\phi)_2$ Euclidean (Quantum) Field Theory, Princeton University Press (1974).
[21] ALBEVERIO, S., HOEGH-KRÖHN, R., Ann. Inst. H. Poincaré A, 1979.
[22] GIELERAK, R., Verification of the Global Markov Property... Preprint FIMETH Zürich 1980. to appear in J. Math. Phys.
[23] ALBEVERIO, S., HOEGH-KRÖHN. R., OLSEN, G., Bielefeld-Marseille, Preprint 1978.
[24] BELLISSARD, J., PICCO, P., 1978, CNRS-CTP preprint.
[25] NELSON, E., J. Funct. Anal. **12** (1973) No 1.
[26] NELSON, E., in Constructive field theory, see ref. [14].
[27] DOBRUSHIN, R. L., MINLOS, R. A., in Problems in Mechanics and Mathematical Physics, Science, Moscow 1976 (in Russian).
[28] MICHLIN, S. G., Course of mathematical physics, "Science", 1968 (in Russian).
[29] ALBEVERIO, S., HOEGH-KRÖHN, R., Commun. Math. Phys. **30** (1973) 171.
[30] ARAKI, H., Math. Phys. **1** (1960) 492.
[31] ALBEVERIO, S., HOEGH-KRÖHN, R., Commun. Math Phys. **30** (1973) 171.
[32] SKRIPNIK, V., Preprint, Institute for Theoretical Physic Kiev. 1972.
[33] SIMON, B., Commun. Math. Phys. **31** (1973) 127.

The journal "Fortschritte der Physik/Progress of Physics" is devoted to the theoretical and experimental study of the fundamental constituents of matter and their interactions, so to elementary particle physics, classical and quantum field theory, theory of gravitation, thermodynamics and statistics, nuclear physics, laser physics, and plasma physics.
The articles should have in general review character.

Manuscripts should be addressed to the Editor-in-Chief of the journal

Prof. Dr. F. Kaschluhn
Sektion Physik der Humboldt-Universität zu Berlin
DDR-1086 Berlin, PSF 1297

or to editors or to members of the Advisory Board.

Terms of subscription for the journal "Fortschritte der Physik/Progress of Physics"

Orders can be sent

- *in the GDR:* to a book-shop or to the Akademie-Verlag, DDR-1086 Berlin, Leipziger Str. 3–4, PF-Nr. 1233;
- *in the other socialist countries:* to a book-shop for foreign language literature or to the competent news-distributing agency;
- *in the FRG and Berlin (West):* to a book-shop or to the wholesale distributing agency Kunst und Wissen, Erich Bieber OHG, Wilhelmstr. 4–6, D-7000 Stuttgart 1;
- *in the other Western European Countries:* to Kunst und Wissen, Erich Bieber GmbH, Dufourstr. 51, CH-8008 Zürich;
- *in other countries:* to the international book- and journal-selling trade, to Buchexport, Volkseigener Außenhandelsbetrieb der DDR, Postfach 160, DDR-7010 Leipzig, or to the Akademie-Verlag, DDR-1086 Berlin, Leipziger Str. 3–4, PF-Nr. 1233.

Zeitschrift „Fortschritte der Physik"

Herausgeber: Prof. Dr. F. Kaschluhn, Prof. A. Lösche, Prof. Dr. R. Rompe.
Verlag: Akademie-Verlag, DDR-1086 Berlin, Leipziger Str. 3–4;
Fernruf: 2236221 und 2236229; Telex-Nr.: 114420; Bank: Staatsbank der DDR, Berlin, Konto-Nr.: 6836-26-20712.

Chefredakteur: Dr. Lutz Rothkirch
Anschrift der Redaktion: Sektion Physik der Humboldt-Universität zu Berlin, DDR-1040 Berlin, Hessische Straße 2.

Veröffentlicht unter der Lizenznummer 1324 des Presseamtes beim Vorsitzenden des Ministerrates der Deutschen Demokratischen Republik.

Gesamtherstellung: VEB Druckhaus „Maxim Gorki",
DDR-7400 Altenburg, Carl-von-Ossietzky-Straße 30/31.

Erscheinungsweise: Die Zeitschrift „Fortschritte der Physik" erscheint monatlich. Die 12 Hefte eines Jahres bilden einen Band. Bezugspreis je Band 216,– M zuzüglich Versandspesen (Preis für die DDR: 120,– M). Preis je Heft 18,– M (Preis für die DDR: 10,– M).

Bestellnummer dieses Heftes: 1027/31/11.

AN(EDV) 57618

Contents of the following issues:

Progress of Physics is indexed
in Current Contents/Physical, Chemical & Earth Sciences.

FSC
www.fsc.org
MIX
Papier | Fördert gute Waldnutzung
FSC® C083411

Kolibri 360 GmbH
Ferdinand-Jühlke-Straße 7
99095 Erfurt, Deutschland
produktsicherheit@kolibri360.de